METHUEN'S MONOGRAPHS
ON BIOLOGICAL SUBJECTS

General Editor: KENNETH MELLANBY, C.B.E.

MULTICOMPARTMENT MODELS
FOR BIOLOGICAL SYSTEMS

Multicompartment Models
for Biological Systems

G. L. ATKINS

Department of Biochemistry
University of Edinburgh Medical School

METHUEN & CO. LTD
11 NEW FETTER LANE LONDON EC4

First published 1969
© *G. L. Atkins 1969*
Printed in Great Britain by
Willmer Brothers Limited, Birkenhead

SBN 416 13820 9

Distributed in the U.S.A.
by Barnes and Noble, Inc.

To Margaret Anne

whose patience and support
has made this possible.

1689

Preface

This book is concerned with the description of certain biological systems in terms of multicompartment models. The systems discussed are those in which the materials present can be considered to be distributed between a number of separate compartments. The rate of transfer of a material out of a given compartment is proportional to the amount or concentration of the material within the compartment. The system can thus be simply described by a set of linear differential equations and this constitutes the model for the system. The analysis of this type of biological system leading to a multicompartment model for its description allows one to obtain quantitative information about the kinetic behaviour of the materials within the system. The generally accepted term for this type of analysis, amongst those of us who study this particular area of biology, is Compartmental Analysis.

I have written this book because I felt that there was a need for an introductory account of the principles, theory, and applications of Compartmental Analysis. At the present time the majority of applications are to mammals by the Medical Sciences e.g. tracer kinetics in physiology and biochemistry, and drug kinetics in pharmacology. I have had in mind the requirements of Honours and Postgraduate students of these particular sciences, but I hope that it will also prove to be useful to those who already have some experience of the techniques and especially those who teach the subject. Many of the ideas and methods described in this book can be applied to other situations and other forms of life by those working in related fields of biology, and I hope that these biologists may also find the book of value.

The original literature is widely dispersed throughout the journals of many scientific disciplines and not all of this is readily accessible to biologists. My approach has been to survey as much as possible of the scattered material and select from it papers which (1) discuss the definitions and assumptions of the subject, (2) provide a theoretical reasoning leading to the mathematical description of some of the simpler systems, (3) are good examples which demonstrate the application of a model to a particular biological situation, (4) discuss the relevance of the results to the original biological system, or (5) discuss the difficulties which exist between describing the system in theoretical terms and fitting experimental data to the mathematical model in order to calculate the constants of the system. The contents follow naturally from this general scheme. After Chapter 1, which is a brief historical introduction, Chapter 2 discusses the basic principles and the terminology of the subject. Chapters 3 and 4 take several simple systems in turn and, for each, describe the theoretical treatment which leads to a mathematical model, illustrate the use of the model in analysing a practical situation, and then give a meaning to the constants calculated from the model. Chapters 5 and 6 describe the techniques which are currently used for fitting experimental data to the types of model presented previously, and for interpreting the mathematical results in terms of biological quantities. The practical limitations of these data fitting techniques are also discussed.

Compartmental Analysis, of necessity, requires some mathematics. In order to make the subject comprehensible and readable to as many kinds of biologists as possible the mathematics has been kept to a minimum. The author makes no apologies for the amount of mathematics presented. The biological sciences are becoming more quantitative, and yet, most biologists have a resistance against mathematics; the sooner this attitude is overcome the better it will be for many branches of biology. Several Appendices have been provided and in these the mathematics of the earlier chapters is explained in more detail for those who are interested. Thus I have illustrated the use of two methods for integrating linear differential equations, and the usefulness of matrices and determinants in one field of biology. The Appendices are meant to be illustrative and not a rigorous treatment, the reader must refer to mathematical or engineering textbooks for this.

I have not attempted to give a comprehensive review of the many applications of Compartmental Analysis. However, for those who may wish to delve deeper into the subject I have tried to give references to the more important papers, reviews and books in the field.

My thanks are due to several of my colleagues and past Honours students in this Department who have read all or parts of the manuscript and have offered much valuable advice and criticism. Any errors or failings which remain must of course be my responsibility. I am also indebted to Miss Di Bennet who most generously typed the initial versions of the manuscript.

Finally, I should like to thank Professor R. B. Fisher for his encouragement and guidance throughout the several stages in the preparation of this book.

G.L.A.

Contents

CHAPTER ONE

Introduction

The origins of the techniques of compartmental analysis can be traced back to the work of Hevesey in which he used radioactive isotopes as tracers. The first application to a biological system utilised radioactive lead (thorium B, ^{212}Pb) to demonstrate the uptake and loss of lead ions by the roots of *Vicia faba* (Hevesey 1923). The first animal experiments were also by Hevesey and his group who measured the metabolism of radioactive bismuth (radium E, ^{210}Bi) in the rabbit (Christiansen, Hevesey, and Lomholt 1924).

After the discovery of deuterium by Urey in 1931, the subsequent efforts of Urey and his colleagues led to the provision of stable isotopes during the later 1930s, and these were used by biologists for metabolic studies. Thus Hevesey and Hofer were the first to use deuterium oxide to study the movement of water in whole animals, firstly in goldfish and then in humans (Hevesey and Hofer 1934a,b).

Schoenheimer and his group used this isotope to prepare deuterated triglycerides and to study triglyceride deposition in mice (Schoenheimer and Rittenberg 1935a,b). Later, as other stable isotopes became available they synthesized many organic compounds involved in animal metabolism—fatty acids, sterols, and amino acids—and thus they were able to investigate the qualitative aspects of the metabolism of lipids and proteins. As a result of this work it was conclusively demonstrated that the constituent compounds of the animal body were constantly being degraded and resynthesized, with only small variations in the amounts present i.e. these compounds were in a dynamic state (Schoenheimer 1940).

1

The first use of artificially prepared radioactive isotopes was that of ^{32}P, also by Hevesey and his colleagues, to measure the overall metabolism of phosphorus firstly in rats and then in humans (Chiewitz and Hevesey 1936, 1937).

The majority of the work using isotopic tracers up to the early 1940s was of a qualitative nature, although a few rough estimates of biological half-lives had been made and some mathematical treatments had been proposed (Artom, Sarzana, and Segre 1938; Burton 1939). The credit for the first adequate quantitative treatment, including suitable definitions of the necessary terminology, goes to the now classic paper of Zilversmit, Entenman, and Fishler (1943). However it must be noted that in the field of drug kinetics, theoretical treatments had been available long before experiments were performed with isotopic tracers e.g. Widmark (1920) and Gehlen (1933). Throughout the 1940s experimental data came to be treated in a more quantitative manner, and with this the concept of compartments and the use of differential equations to describe the behaviour of tracers became increasingly more important. Eventually this led to the definition of more precise terminology and to the development of general mathematical treatments for analyzing experimental data. Concerning the latter, the more important papers are those of Sheppard (1948), Sheppard and Householder (1951), and Berman and Schoenfeld (1956).

The more recent advances in compartmental analysis have involved the introduction of computers. First the use of electric analogue computers for simulation and curve fitting (e.g. Hickey and Brownell 1954), then the use of digital computers for curve fitting (Worsley and Lax 1962; Berman, Shahn, and Weiss 1962), and lastly the use of digital computers for the simulation of very complex systems (e.g. Garfinkel 1963, 1966).

The advent of these machines in conjunction with the increased quantity and quality of experimental data has allowed systems of increasing complexity to be studied. The present developments of compartmental analysis seem to lie partly in developing more precise mathematical techniques for interpreting experimental data (e.g. operator methods, transform methods), partly in the study of complex systems in great detail (e.g. iodine and iron metabolism), and partly in extending the theory to include non-linear systems (e.g. hormonal control mechanisms).

The theory of compartments and the use of differential equations is not the only technique that has been developed for the interpretation of tracer and allied experiments. Theories involving integral equations, the use of network analysis, and a probabilistic approach have all been described, but none of these has enjoyed the popularity of compartmental analysis. For this reason, this book will be concerned only with compartmental analysis. A list of the more important papers concerned with the other techniques has been given for those who may wish to study them.

REFERENCES

ARTOM, C., SARZANA, G. and SEGRE, E. (1938). *Archs int. Physiol.*, **47**, 245. Influence des graisses alimentaires sur la formation des phospholipides dans les tissus animaux.

BERMAN, M. and SCHOENFELD, R. (1956). *J. appl. Phys.*, **27**, 1361. Invariants in experimental data on linear kinetics and the formulation of models.

BERMAN, M., SHAHN, E. and WEISS, M. F. (1962). *Biophys. J.*, **2**, 275. The routine fitting of kinetic data to models: a mathematical formalism for digital computers.

BURTON, A. C. (1939). *J. cell. comp. Physiol.*, **14**, 327. The properties of the steady state compared to those of equilibrium as shown in characteristic biological behaviour.

CHIEWITZ, O. and HEVESEY, G. (1935). *Nature, Lond.*, **136**, 754. Radioactive indicators in the study of phosphorus metabolism in rats.

CHIEWITZ, O. and HEVESEY, G. (1937). *Biol. Meddr.*, **13** (9), 1. Studies on the metabolism of phosphorus in animals.

CHRISTIANSEN, I. A., HEVESEY, G. and LOMHOLT, S. (1924). *C. r. hebd. Séanc. Acad. Sci., Paris*, **178**, 1324. Recherches, par une méthode radiochimique, sur la circulation du bismuth dans l'organisme.

GARFINKEL, D. (1963). *Ann. N. Y. Acad. Sci.*, **108**, 293. Digital computer simulation of systems apparently compartmented at the cellular level.

GARFINKEL, D. (1966). *Biochem. Soc. Symp.*, **26**, 81. The digital computer as a biochemical instrument: simulation of multi-enzyme systems.

B

GEHLEN, W. (1933). *Arch. exp. Path. Pharmak.*, **171**, 541. Wirkungs-stärke intravenös verabreichter Arzneimittel als Zeitfunktion.

HEVESEY, G. (1923). *Biochem. J.*, **17**, 439. The absorption and trans-location of lead by plants.

HEVESEY, G. and HOFER, E. (1934a). *Hoppe-Seyler's Z. physiol. Chem.*, **225**, 28. Der Austach des Wassers im Fischkörper.

HEVESEY, G. and HOFER, E. (1934b). *Klin. Wschr.*, **13**, 1524. Die Verweilzeit des Wassers im menschlichen Körper, untersucht mit Hilfe von 'schweren' Wasser als Indicator.

HICKEY, F. C. and BROWNELL, G. L. (1954). *J. clin. Endocr. Metab.*, **14**, 1423. Dynamic analysis of iodine metabolism in four normal subjects.

SCHOENHEIMER, R. (1940). 'The Dynamic State of Body Constituents'. (Harvard University Press: Cambridge, Mass.)

SCHOENHEIMER, R. and RITTENBERG, D. (1935a). *J. biol. Chem.*, **111**, 163. Deuterium as an indicator in the study of intermediary metab-olism. I.

SCHOENHEIMER, R. and RITTENBERG, D. (1935b). *J. biol. Chem.*, **111**, 175. Deuterium as an indicator in the study of intermediary metab-olism. III. The rôle of the fat tissues.

SHEPPARD, C. W. (1948). *J. appl. Phys.*, **19**, 70. The theory of the study of transfers within a multi-compartment system using isotopic tracers.

SHEPPARD, C. W. and HOUSEHOLDER, A. S. (1951). *J. appl. Phys.*, **22**, 510. The mathematical basis of the interpretation of tracer experi-ments in closed steady state systems.

WIDMARK, E. M. P. (1920). *Acta med. scand.*, **52**, 87. Studies in the concentration of indifferent narcotics in blood and tissues.

WORSLEY, B. H. and LAX, L. C. (1962). *Biochim. biophys. Acta*, **59**, 1. Selection of a numerical technique for analyzing experimental data of the decay type with special reference to the use of tracers in biological systems.

ZILVERSMIT, D. B., ENTENMAN, C. and FISHLER, M. C. (1943). *J. gen. Physiol.*, **26**, 325. On the calculation of turnover time and turnover rate from experiments involving the use of labeling agents.

Integral Equations
BECK, J. S. and RESCIGNO, A. (1964). *J. theor. Biol.*, **6**, 1. Determination of precursor order and particular weighting functions from kinetic data.

BRANSON, H. (1946). *Bull. math. Biophys.*, **8**, 159. A mathematical description of metabolising systems: I.

BRANSON, H. (1947). *Bull. math. Biophys.*, **9**, 93. A mathematical description of metabolising systems: II.

BRANSON, H. (1947). *Science, N.Y.*, **106**, 404. The use of isotopes to determine the rate of a biochemical reaction.

BRANSON, H. (1948). *Cold Spring Harb. Symp. quant. Biol.*, **13**, 35. The use of isotopes in an integral equation description of metabolising systems.

BRANSON, H. (1952). *Archs. Biochem. Biophys.*, **36**, 48. The kinetics of reactions in biological systems.

BRANSON, H. (1952). *Archs Biochem. Biophys.*, **36**, 60. Metabolic pathways from tracer experiments.

BRANSON, H. (1963). *Ann. N.Y. Acad. Sci.*, **108**, 4. The integral equation representation of reactions in compartment systems.

HART, H. E. (1963). *Ann. N.Y. Acad. Sci.*, **108**, 23. An integral equation formulation of perturbation in tracer analysis.

HART, H. E. (1966). *Bull. math. Biophys.*, **28**, 261. Analysis of tracer experiments: VII. General multicompartment systems imbedded in non-homogeneous inaccessible media.

HEARON, J. Z. (1953), *Bull. math. Biophys.*, **15**, 269. A note on the integral equation description of metabolising systems.

STEPHENSON, J. L. (1960). *Bull. math. Biophys.*, **22**, 1. Theory of transport in linear biological systems: I. Fundamental integral equation.

WIJSMAN, R. A. (1953). *Bull. math. Biophys.*, **15**, 261. A critical investigation of the integral description of metabolising systems.

Network Analysis

RESCIGNO, A. (1960). *Biochim. biophys. Acta*, **37**, 463. Synthesis of a multicompartment biological model.

RESCIGNO, A. and SEGRE, G. (1961). *J. theor. Biol.*, **1**, 498. The precursor-product relationship.

RESCIGNO, A. and SEGRE, G. (1962). *J. theor. Biol.*, **3**, 149. Analysis of multicompartmented biological systems.

RESCIGNO, A. (1963). *Ann. N.Y. Acad. Sci.*, **108**, 204. Flow diagrams of multi-compartment systems.

RESCIGNO, A. (1964). *Bull. math. Biophys.*, **26**, 31. On some topological properties of the systems of compartments.

SCHOENFELD, R. L. (1963). *Ann. N.Y. Acad. Sci.*, **108**, 69. Linear network theory and tracer analysis.

Stochastic Methods

BERGNER, P. E. E. (1961). *J. theor. Biol.*, **1**, 120. Tracer dynamics: I. A tentative approach and definition of fundamental concepts.

BERGNER, P. E. E. (1961). *J. theor. Biol.*, **1**, 359. Tracer dynamics: II. The limiting properties of the tracer systems.

BERGNER, P. E. E. (1962). *Acta radiol.*, **S.210**, 1. The significance of certain tracer kinetical methods, especially with respect to the tracer dynamic definition of metabolic turnover.

BERGNER, P. E. E. (1964). *J. theor. Biol.*, 6, 137. Tracer dynamics and the determination of pool-sizes and turnover factors in metabolic systems.

CHAPTER TWO

The Terminology of Compartmental Analysis

2.1 Introduction

Many terms are used and have been used in compartmental analysis. Often there has been controversy as to the correct definition of some of these terms, and on many occasions, including some of the review articles in the field, the terms are not defined at all. As a prerequisite to the further study of compartmental analysis this chapter is concerned with discussing and defining the terminology, and in the case of the steady state, discussing the assumptions involved in the use of tracers to study this state.

2.2 Systems

A *mother substance* (Bergner 1961), *basic substance* (Sharney *et al.* 1965), or often simply just *substance* (Zilversmit, Entenman, and Fishler 1943) is the chemical compound, or chemical group, or particular atom whose quantitative behaviour it is desired to study. For example, the substance could be an inorganic ion (Na^+), a normal metabolite (cholesterol), a hormone (aldosterone), a drug (aspirin), a particular chemical group (methyl-), or one individual atom (nitrogen during protein metabolism).

A *system* is any biochemical or physiological system within which it is desired to study the behaviour of a substance, and on which experiments can be performed. Systems can be whole animals or plants, surviving organs or parts of animals or plants, living cells of animals, plants or bacteria, or the subcellular components of cells e.g. nuclei or mitochondria. The system studied may be only part of an intact organism. An example is the study of the disappearance of ^{131}I from the plasma of an animal; the system being studied here

7

consists simply of the transport of iodine to and from the blood plasma.

A *closed system* is a system which no material can either enter or leave. An *in vitro* study of the exchange of Na^+ or K^+ between erythrocytes and plasma or compatible solutions would constitute a closed system. An *open system* is any system which can exchange materials with its environment. Some systems, for example perfusion of surviving organs, can be either open systems or closed systems depending on whether the perfusate is recycled or not, or whether an important metabolite such as carbon dioxide is lost from the system.

2.3 Models

The concepts of a *metabolic pool* and a *compartment* are often confused, but there is a clear distinction between the two.

The original idea of a *metabolic pool* was first discussed by Schoenheimer (1940), and was later defined by Sprinson and Rittenberg (1949) as follows: 'The metabolic pool of the animal (or organ or cell) is defined as that mixture of compounds derived either from the diet or from the breakdown of the tissues, which the animal (or organ or cell) employs for the synthesis of tissue constituents.'

These authors seemed to consider that one could lump together all the amino acids, and possibly ammonia and peptides, and regard them as kinetically equivalent sources not only for the synthesis of protein but also of every other important nitrogen-containing compound of the cell. There is an inherent vagueness in the use of the term 'pool', therefore its use ought to be restricted to those occasions when one is considering in a qualitative fashion the sources from which cell materials are synthesized.

On the other hand, a *compartment*, a term first used by Sheppard (1948), is a quantity of a substance which has uniform and distinguishable kinetics of transformation or transport. Thus, one can talk about the pool of α-amino-nitrogen in the body, but, because the interchange of different amino acids, with one another or with different synthetic or degradative pathways, proceeds with different kinetics, the pool must be considered to be distributed between many compartments. The size of a compartment is determined by the amount of substance within the compartment and is in units of mass i.e. moles, m-moles, or grammes.

The transfer of a substance from one compartment to another may represent either the transport of the substance from one physiological location to another, or the transformation of the substance from one chemical form within the same physiological boundaries. Thus, the α-amino-nitrogen of glycine originally present in the liver may either be transported within intact glycine molecules to α-amino-nitrogen of glycine in the plasma, or it may be transformed via an enzymic reaction into the α-amino-nitrogen of serine and remain within the liver. Schoenheimer and Rittenberg (1935) were the first to mention the distinction between transport and conversion.

A *theoretical model* is the representation of the kinetics of a substance in a biological system. The idea of compartmentalisation is fundamental to the construction of these theoretical models. The kinetics can therefore be devised in which transfer occurs between compartments in such a way that some are transfers between compartments of the same metabolic pool and some involve transformations into elements of different pools.

A *mathematical model* is a set of equations which are derived from the theoretical model and describe the concentrations and amounts of the substance involved as functions of time. As an

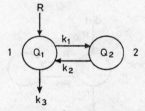

Fig. 2.1 A theoretical model, which contains two compartments, to represent the kinetics of inorganic phosphate in an animal. Q_1 and Q_2 are the amounts of phosphate in the two compartments, and the arrows represent flows of phosphate into and out of the compartments.

example, the kinetics of inorganic phosphate in an animal over a certain time interval can be represented by the theoretical model of Figure 2.1. Here 1 and 2 are two compartments, Q_1 and Q_2 are the amounts of phosphate in the two compartments, and R is the flow rate into compartment 1. It is assumed that transfer out of each compartment is proportional to the concentration in the compartment, the constants of proportionality, k_1, k_2, and k_3, are called the rate constants. The mathematical model corresponding to this theoretical model is:

$$\frac{dQ_1}{dt} = R + k_2 \cdot Q_2 - k_1 \cdot Q_1 - k_3 \cdot Q_1$$

$$\frac{dQ_2}{dt} = k_1 \cdot Q_1 - k_2 \cdot Q_2$$

Compartmental analysis comprises the various techniques which are used in order to describe the behaviour of a biological system in terms of a theoretical or mathematical model. Very often several theoretical models will give rise to the same mathematical model, so that although a mathematical model will describe a system it is often impossible to decide which theoretical model applies. To illustrate this possibility one can consider the experiments of Sprinson and Rittenberg (1949) and Rittenberg (1951). Both papers are concerned with the synthesis of protein from an organic nitrogen compartment and with the synthesis of urea. The two models are shown in Figures 2.2 and 2.3, respectively. In both

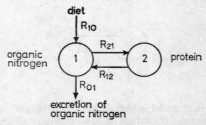

Fig. 2.2 The model of Sprinson and Rittenberg (1949) to represent the metabolism of organic nitrogen in man.

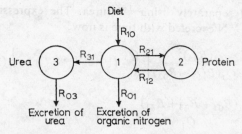

Fig. 2.3 The model of Rittenberg (1951) to represent the metabolism of organic nitrogen in man. A separate compartment for urea is provided.

experiments, ^{15}N-glycine was taken orally by human subjects and then the excretion of ^{15}N was measured over approximately 50–70 hr. Both models are simplified in that no ^{15}N is assumed to be returned to the organic nitrogen compartment during the period of the experiment by protein degradation i.e. $R_{12} = 0$.

In the earlier model (Fig. 2.2), urea is considered to be synthesized and excreted directly i.e. there is no urea compartment. The ^{15}N-enrichment of total urinary nitrogen compounds was measured and the data, calculated as fraction of label excreted at time t, was given by

$$\text{fraction} = \frac{R_{01}}{(R_{21} + R_{01})} \left[1 - e^{-(R_{01} + R_{21}/Q_1) \cdot t} \right]$$

From the results, the following information was calculated:

rate of excretion of organic nitrogen $= R_{01} \approx 13\text{g } N/\text{day}$

rate of protein synthesis $= R_{21} \approx 14\text{g } N/\text{day}$

amount of organic nitrogen in body $= Q_1 \approx 35\text{g } N$

In the later model (Fig. 2.3) an additional compartment was added in order to separate urea from the remainder of the organic nitrogen compounds. The parameters for this compartment were determined separately using ^{15}N-urea. The expression relating fraction of ^{15}N excreted with time is now:

$$\text{fraction} = \frac{(R_{03})^2}{Q_1(BQ_3 - R_{03})}\left[\frac{Q_3}{R_{03}}(1 - e^{-(R_{03}/Q_3).t}) - \frac{1}{B}(1 - e^{-B.t})\right]$$

$$(2.1)$$

where $B = \dfrac{(R_{01} + R_{03} + R_{21})}{Q_1}$

Some of the values calculated in this series of experiments were:

rate of excretion of organic nitrogen	$= (R_{01} + R_{03})$	$\approx 11.5\text{g } N/\text{day}$
rate of protein synthesis	$= R_{21}$	$\approx 39\text{g } N/\text{day}$
amount of organic nitrogen in body	$= Q_1$	$\approx 0.61\text{g } N$
amount of urea in body	$= Q_3$	$\approx 5.65\text{g } N$

Note that the values of $Q_3/R_{03} \approx 0.631$ day, $R_{03}/Q_3 \approx 1.58$ day^{-1}, $1/B \approx 0.012$ day, and $B \approx 82.8$ day^{-1}. Therefore, equation 2.1 over the time of the experiment approximates to

$$\text{fraction} = \frac{R_{03}\,Q_3}{Q_1(BQ_3 - R_{03})}(1 - e^{-(R_{03}/Q_3).t})$$

If one was not aware of the third (urea) compartment the data from the experiments of Rittenberg would also fit the simpler equation of the type produced by Sprinson and Rittenberg. Thus, by only considering the labelling of organic nitrogen compounds in the urine, it is not possible to choose between the two models. In addition, the values calculated according to the two theoretical models are quite different.

More detailed discussions on the use of models in biology and medicine may be found in *Symp. Soc. exp. Biol.* (1960) and Nooney (1966).

2.4 Concept of turnover

The concept of the *steady state* is fundamental to a great deal, but by no means all, of compartmental analysis. A steady state exists in a system of a mixture of substances if they are transported from one part of the system to another, or transformed from one into another, and yet, because their rates of removal are equal to their rates of replacement, their concentrations in all relevant compartments remain constant during the interval over which observations are made. In the case of a closed system the steady state is also a state of *dynamic equilibrium*, but as soon as the system becomes open the equilibrium disappears. Thus the term 'dynamic equilibrium' applies only to closed systems. 'Steady state', however, could apply to either system, but by convention its use is restricted to open systems. Burton (1939) discusses fully the concept of steady state in biological systems especially in relation to the thermodynamic equilibrium state. If the steady state exists only during the time of an experiment it is occasionally called a *quasi-steady state*. Since most steady states are present in biological systems for only a finite time the distinction is unnecessary and the term 'steady state' will be used throughout this book.

That animals are in a steady state was first demonstrated by the work of Schoenheimer and his colleagues (Schoenheimer 1940), but the concept seems to have been accepted by earlier biologists (e.g. Hill 1930) and was strongly indicated by the work of Borsook and Keighley (1935) on protein metabolism and nitrogen balance.

When used qualitatively, *turnover* refers to the transference or transformation of a given substance between metabolic pools or compartments (Zilversmit, Entenman, and Fishler 1943). *Exchange* is a special form of turnover in which there is a one-for-one substitution of molecules of a substance for those of another i.e. there is simultaneous and equal transfer in and out of metabolic pools or compartments.

The quantitative definition of turnover when applied to systems in a steady state has produced both confusion and controversy. One definition has the unit of time^{-1} i.e. amount of substance turned over per unit amount of substance per unit time, and has been variously called the turnover rate (Feller, Strisower, and Chaikoff 1950; Siri 1949; Kleiber 1955; Mawson 1955), the transfer coefficient (Solomon 1953), the relative turnover rate (Reiner 1953), the fractional turnover rate (Zilversmit 1955), the flow constant (Bergner 1959), and the rate constant (Robertson 1957). The second definition has the unit of mass . time^{-1} i.e. amount of substance turned over per unit of time, and has been called the turnover rate (Zilversmit, Entenman, and Fishler 1943; Zilversmit 1955; Robertson 1957), the flux (Solomon 1953), and the flux rate (Mawson 1955). Thus the term turnover rate has been, and often still is, used for different units, and this is the source of the confusion. The terminology proposed by Robertson is probably the more exact physiologically and finds favour amongst the majority of users of compartmental analysis. He suggests, and the author is in agreement, that *rate constant* is used for units in time^{-1}, and *turnover rate* for units in mass . time^{-1}. It would seem wise, therefore, when reading a published paper to be sure what the author means by turnover rate, and also to define carefully one's own meaning for the term.

Half-life as a measure of turnover is easily understood and is therefore very widely used. It originated in the study of radioactive disintegration where it denotes the time required for the disintegration of half of the radioactive atoms originally present in a sample. Its first biological application was to a non-radioactive experiment: Hevesey and Hofer (1934) applied it to the turnover of deuterium oxide in men. Its application is based on the use of similar first order equations in radioactive disintegrations, chemical kinetics, and compartmental analysis of steady state systems. For those systems which can be represented by the mathematical model $dQ/dt = -kQ$, then the half-life, usually symbolised by $t_{\frac{1}{2}}$, can be calculated from the rate constant k by $t_{\frac{1}{2}} = \log_e 2/k = 0.693/k$. It should be noted that with more complex systems where exponential constants λ are determined from experimental data (see Chapter 3), the quantity $0.693/\lambda$ has no physiological significance.

Turnover time is likewise related to the rate constants applicable to the substance concerned, and is usually defined as the time required for the appearance or disappearance of an amount of the substance present in the pool or compartment (Zilversmit, Entenman, and Fishler, 1943). This definition has been critically discussed by Reiner (1953). In the simple mathematical model above, the turnover time would be given by $1/k$. Perhaps a more accurate description for this term would be that suggested by Mawson (1955) as the average life of a molecule of a substance within a compartment. This description was first used by Hevesey and Hofer (1934).

2.5 Tracers

Direct chemical analysis of a system will only yield information concerning the static condition of the system. Generally, results are limited to the measurement of concentrations of substances within various parts of the system or to measurement of the amounts of substances taken in or excreted by the system as a whole. Little information is obtained about the dynamic behaviour of the substance within the system. Therefore, in order to study the kinetics of a substance within a system it is necessary to label the substance. In compartmental analysis these labels are called *tracers*, and to be useful they must be easily detected by an observer. Tracers should not interfere with the behaviour of the substance being studied, and their kinetics should be equivalent to that of the mother substance.

A tracer may be an isotopic form of an element and be either radioactive or stable. A stable isotope used as a tracer is usually one which is naturally present in only very small amounts e.g. ^{13}C or ^{18}O. The vast majority of tracers are isotopes. Certain metabolically inert and rapidly diffusible compounds have been used as a tracer for body water e.g. N-acetyl-4-amino-antipyrene. Coloured or fluorescent dyes which can bind to metabolised materials are occasionally used e.g. T-1824, a blue dye used to trace plasma proteins.

A *perfect tracer* has the following properties—
1. The biological system should be unable to distinguish between

the mother substance and its tracer. The tracer therefore undergoes the same metabolic changes as the mother substance (Schoenheimer and Rittenberg 1935).

2. A tracer can be added in small amounts and therefore does not disturb the steady state existing in the system as a whole.

3. A tracer when initially added to a system is not in equilibrium and its quantitative changes can be mathematically analyzed and described as a function of time. These changes reflect the rates of transfer and transformation of the substance under investigation.

4. If the tracer is an isotope, there should be no exchange of the isotope between the labelled compound and other compounds e.g. deuterium should not be bound to oxygen or nitrogen atoms.

The quantitative changes of the tracer are usually measured in units of concentration within the mother substance. For tracers which are dyes, or compounds not normally present in the biological system, this is quite straightforward. Radioactive tracers are measured by virtue of their radioactivity using counting devices which are generally of unknown but constant efficiency. Their concentration or *specific activity* is thus expressed in counting rates per m-mole, counts per min. per mg, or occasionally as μc per m-mole. The *relative specific activity* of a traced substance is the ratio of the specific activity of a substance at a given time to the specific activity of either the same substance at a different time or of a different substance at the same time.

Concentrations of stable isotopes are usually measured in a mass spectrometer, or occasionally by density measurements, and are expressed as the ratio of the number of atoms of tracer isotope to the number of atoms of the most abundant natural isotope. This is called the *abundance ratio*, or, if expressed as a percentage, the *atom per cent*. Sometimes the tracer isotope is also a natural isotope, in which case the natural concentration is subtracted from the experimental concentration and the result expressed as *atom per cent excess*. This procedure depends on the experimental fact that naturally occurring isotopes are nearly always present in constant abundance ratios regardless of the source. For example, measurements of the density of water (an estimation of the deuterium content) obtained either from natural sources or from various animal tissues show differences of less than three parts per million (Stewart and Holcomb 1934; Breusch and Hofer 1934).

Schoenheimer and Rittenberg (1939) determined the normal abundance of ^{15}N in several amino acids which had been obtained from natural sources, and for all except arginine they were close to the normal value of 0·368 atom per cent. The value for arginine, however, was about 0·010 atom per cent higher, and this is presumably due to repeated fractionation of the isotopes of nitrogen by recycling in the urea cycle. Care has to be taken, therefore, that a similar process may not occur in an experiment.

The above are the most common terms used in compartmental analysis. Other terms do occur in special uses of the technique, and these will be defined or discussed as required.

REFERENCES

BERGNER, P. E. E. (1959). *Expl Cell Res.*, **17**, 328. Dynamic aspects of a method in tracer kinetics.

BERGNER, P. E. E. (1961). *J. theor. Biol.*, **1**, 120. Tracer dynamics: I. A tentative approach and definition of fundamental concepts.

BORSOOK, H. and KEIGHLEY, G. L. (1935). *Proc. R. Soc.*, **118B**, 488. The 'continuing' metabolism of nitrogen in animals.

BREUSH, F. and HOFER, E. (1934). *Klin. Wschr.*, **13**, 1815. Uber das Verhältnis des schweren Wassers zum Leichten im Organismus.

BURTON, A. C. (1939). *J. cell. comp. Physiol.*, **14**, 327. The properties of the steady state compared to those of equilibrium as shown in characteristic biological behaviour.

FELLER, D. D., STRISOWER, E. H. and CHAIKOFF, I. L. (1950). *J. biol. Chem.*, **187**, 571. Turnover and oxidation of body glucose in normal and alloxan-diabetic rats.

HEVESEY, G. and HOFER, E. (1934). *Nature, Lond.*, **134**, 879. Elimination of water from the human body.

HILL, A. V. (1930). *Trans. Faraday Soc.*, **26**, 667. Membrane phenomena in living matter: equilibrium or steady state.

KLEIBER, M. (1955). *Nature, Lond.*, **175**, 342. Meaning of 'turnover' in biochemistry.

MAWSON, C. A. (1955). *Nature, Lond.*, **176**, 317. Meaning of 'turnover' in biochemistry.

NOONEY, G. C. (1966). *J. chron. Dis.*, **19**, 325. Mathematical models in medicine: a diagnosis.

REINER, J. M. (1953). *Archs Biochem. Biophys.*, **46**, 53. The study of metabolic turnover rates by means of isotopic tracers. I. Fundamental relations.

RITTENBERG, D. (1951). A method for the evaluation of the rate of protein synthesis in man. *In* Ciba Foundation Conference, Isotopes in Biochemistry, p. 190.

ROBERTSON, J. S. (1957). *Physiol. Rev.*, **37**, 133. Theory and use of tracers in determining transfer rates in biological systems.

SCHOENHEIMER, R. (1940). 'The Dynamic State of Body Constituents'. (Harvard University Press: Cambridge, Mass.)

SCHOENHEIMER, R. and RITTENBERG, D. (1935). *J. biol. Chem.*, **111**, 163. Deuterium as an indicator in the study of intermediary metabolism. I.

SCHOENHEIMER, R. and RITTENBERG, D. (1939). *J. biol. Chem.*, **127**, 285. General considerations in the application of isotopes to the study of protein metabolism. The normal abundance of nitrogen isotopes in amino acids.

SHARNEY, L., WASSERMAN, L. R., GEVIRTZ, N. R., SCHWARTZ, L. and TENDLER, D. (1965). *Am. J. med. Electron.*, **4**, 95. Significance of the time lag in 'tracer' movement: representation of unidirectionally connected pool sequences by time lag.

SHEPPARD, C. W. (1948). *J. appl. Phys.*, **19**, 70. The theory of the study of transfers within a multi-compartment system using isotopic tracers.

SIRI, W. E. (1949). 'Isotopic Tracers and Nuclear Radiations'. (McGraw-Hill: New York.)

SOLOMON, A. K. (1953). *Adv. biol. med. Phys.*, **3**, 65. The kinetics of biological processes. Special problems connected with the use of tracers.

SPRINSON, D. B. and RITTENBERG, D. (1949). *J. biol. Chem.*, **180**, 715. The rate of interaction of the amino acids of the diet with the tissue proteins.

STEWART, W. W. and HOLCOMB, R. (1934). *J. Am. chem. Soc.*, **56**, 1422. The biological separation of heavy water.

Symp. Soc. exp. Biol. (1960). **14**. Models and analogues in biology.

ZILVERSMIT, D. B. (1955). *Nature, Lond.*, **175**, 863. Meaning of turnover in biochemistry.

ZILVERSMIT, D. B., ENTENMAN, C. and FISHLER, M. C. (1943). *J. gen. Physiol.*, **26**, 325. On the calculation of turnover time and turnover rate from experiments involving the use of labeling agents.

CHAPTER THREE

Systems in a Steady State

3.1 Introduction

In this chapter we will consider the theoretical aspects of systems in a steady state. Because they are in a steady state the turnover rates between compartments must be constant, and also the amounts of substances in each compartment must be constant. In these systems the turnover rate is the product of the amount of substance in a compartment and the relevant rate constant. Hence, using the concepts of chemical kinetics, the transfer and transformation processes are comparable to first order reactions, and the mathematical equations which describe these processes are first order linear differential equations. Systems which are in a steady state are therefore also described as linear systems (Zilversmit, Entenman, and Fishler 1943; Solomon 1953).

When the technique of compartmental analysis is applied to the study of steady state systems, several assumptions or approximations are usually made in order that the mathematical models may be made as simple as possible. Firstly, the kinetic processes are assumed to be irreversible. Secondly, the molecules (or atoms) of the substances in the various compartments participate in the processes in a random fashion i.e. there is no distinction between old and newly formed molecules. Thirdly, within the compartments instantaneous and homogeneous mixing of the molecules occurs (Zilversmit, Entenman, and Fishler 1943). In fact, there is a finite time of mixing, but this approximates to instantaneous mixing provided that the time of mixing is very small compared with the turnover time for the substance. Fourthly, tracers are assumed to be

19

perfect tracers as defined in Chapter 2. Lastly, the system must be irreducible i.e. it must be impossible to represent the behaviour of tracers within the system by using fewer compartments (Bergner 1965). In practice it is found that the above assumptions or approximations are valid for the vast majority of systems studied, but where an approximation is found not to apply then suitable mathematical techniques can be used to circumvent the difficulty. Some of these techniques will be described later.

The general procedure which is followed in order to study the behaviour of a system using tracers is as follows—

1. A known amount of tracer substance is introduced into one of the compartments of the system.
2. At suitable subsequent time intervals, samples are taken from this and/or other compartments and the specific activities of the substances are determined.
3. Using the theoretical and mathematical models of the system, the experimental data are analyzed in order to calculate the various parameters of the model.
4. If the fit of the data to the model is not good, then further models must be devised.

The present chapter is concerned with discussing the theoretical treatments which are applicable to certain simple systems.

3.2 Symbols used

The symbols used by different authors vary considerably. The set chosen for this book follows closely a set suggested in a report to the International Commission on Radiation Units by Brownell, Berman, and Robertson (1968), and corresponds approximately to the most general usage. The only symbol which the present author disagrees with is the use of a_j for specific activity and a_{jn} for the coefficient for an exponential term. X_i is suggested as a better symbol for the latter quantity. In order to illustrate the present confusion Table 3.1 reproduces several alternative sets which have been used. It will be noted that no two authors are in agreement!

TABLE 3.1

Variation between authors in the symbols used for the parameters of compartmental analysis.

Author	Amount of substance	Amount of tracer	Specific activity	Rate constant	Turnover rate	Exponential coefficient	Exponential constant
This book	Q_i	q_i	a_i	k_{ij}	R_{ij}	X_i	λ_i
Berman and Schoenfeld (1956)	C_i	q_i	v_i	λ_{ij}	$\lambda_{ji}q_i$	A_i, a_i	λ_i
Hart (1955)	—	x_j	—	a_{ij}, k	—	A_i	λ_i
Reiner (1953)	$A, B, \ldots$	$A^*, B^*, \ldots$	$S_a, S_b, \ldots$	K_i	—	—	—
Rescigno (1954, 1956)	a_i	X_i	x_i	k_{ij}	v_i	C_i	λ_i
Robertson (1957)	S_i	S_i	—	k_{ij}	ρ_{ij}	a_{ij}	λ_i
Sharney et al. (1963)	—	q_i	a_j	—	r_{ij}	x_i	λ_i
Sheppard (1948)	S_i	S_i	—	—	p_{ii}, dS_{ij}	—	—
Sheppard and Householder (1951)	S_i	S_i	a_i	—	p_{ii}, dS_{ij}	x_i	λ_i
Sheppard (1962)	S_i	S_i	a_i	k_{ij}	ρ_{ii}	X_i	λ_i
Solomon (1949, 1953)	—	$P, q, \ldots$	—	k_{ij}	—	—	—
Solomon (1960)	S_i	P_i	P^*_i	k_{ij}	Φ_{ij}	C_i	λ

t	time
n	number of compartments in a system
Q_j	amount of substance in compartment j (mass)
q_j	amount of tracer in compartment j (mass)
$\dfrac{dq_j}{dt}$	rate of change of q_j with respect to time (mass . time^{-1})
$a_j = \dfrac{q_j}{C_j}$	abundance ratio for a stable isotope, or specific activity for a radioactive tracer, in compartment j
$\dfrac{a_j}{a_j(0)}$	relative specific activity of tracer in compartment j
k_{ij}	rate constant for transport into compartment i from compartment j (time^{-1})
R_{ij}	rate of transport of unlabelled substance into compartment i from compartment j (mass . time^{-1})
$q_j(0)$	the amount of tracer in compartment j when $t = 0$ (mass)
$a_j(0)$	the specific activity of the tracer in compartment j when t = 0.
X_i	coefficient of the ith exponential term
λ_i	constant of the ith exponential term (time^{-1})

Note that the rate constant k_{ij} applies to transport into compartment i from compartment j. This is in accord with the notation used in all the important theoretical papers concerning compartmental analysis, although logically one would expect the symbol to represent the reverse process: transport from compartment i into compartment j. Although the symbol given here is preferable, either representation is valid in practice, provided that a clear definition is given.

The abundance ratio of a tracer, a_j, is given by

$$a_j = \frac{(\text{amount of tracer in compartment } j)}{(\text{amount of substance in compartment } j)} = \frac{q_j}{Q_j}$$

hence $q_j = a_j Q_j$

If the tracer is radioactive and the efficiency of the counting instrument used to measure the specific activities is E, then

specific activity = (counting efficiency) × (abundance ratio)

i.e.
$$a_j = E . \frac{q_j}{Q_j}$$

hence
$$q_j = \frac{1}{E} . a_j Q_j$$

The rate of change of the amount of tracer is $d(q_j)/dt$

Thus,
$$\frac{d(q_j)}{dt} = \frac{d(a_j)}{dt} . Q_j \qquad \text{(stable tracer) (3.1)}$$

or,
$$\frac{d(q_j)}{dt} = \frac{1}{E} . \frac{d(a_j)}{dt} . Q_j \qquad \text{(radioactive tracer)}$$

For all determinations of specific activity within an experiment E should be a constant. Inspection of the differential equations which describe the mathematical models in the following pages shows that $1/E$ can be divided out, and thus $d(q_j)/dt$ can be safely replaced by $[d(a_j)/dt] . Q_j$ for all types of tracer. It is not uncommon to use specific activity when abundance ratio is the valid term i.e. with stable isotopes or with non-isotopic tracers such as Evan's Blue. There is no serious objection to this, again because $1/E$ can be divided out of the differential equations.

It will be noted that the equations in the subsequent procedures have been solved in terms of specific activities instead of total activities. The reason is that most data obtained from the use of tracers are in terms of specific activity. However, if the data are in the form of quantity of tracer it is simple to correct the following equations by substituting q/Q for a.

3.3 Isotopic dilution

Let the compartment of Figure 3.1 contain an amount Q_1 of unlabelled substance. Add an amount q_1 of tracer whose specific activity is $a_1(0)$. Allow time for complete mixing of the tracer with

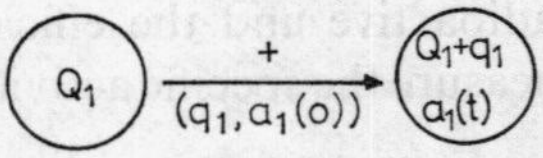

$$\boxed{Q_1} \xrightarrow[\substack{+ \\ (q_1, a_1(o))}]{} \boxed{\substack{Q_1+q_1 \\ a_1(t)}}$$

Fig. 3.1 Isotopic dilution. An amount q_1 of tracer with specific activity $a_1(0)$ is added to an amount Q_1 of unlabelled substance, the resultant specific activity becomes $a_1(t)$.

the contents of the compartment. Remove a sample from the compartment and isolate from it by any appropriate purification procedure a specimen of the substance. Determine the specific activity $a_1(t)$ of this specimen. The amount of radioactivity originally in the tracer is $q_1 . a_1(0)$ and is the same as that present in the compartment after mixing i.e. $(Q_1 + q_1) . a_1(t)$.

Thus,

$$q_1 a_1(0) = (Q_1 + q_1) a_1(t)$$

Rearranging,

$$Q_1 = q_1 \left(\frac{a_1(0)}{a_1(t)} - 1 \right) \tag{3.2}$$

In those determinations where $Q_1 \gg q_1$ equation 3.2 becomes

$$Q_1 = q_1 . \frac{a_1(0)}{a_1(t)} \tag{3.3}$$

Thus the amount of material originally present in the compartment can be calculated.

Rittenberg and Foster (1940) used this method for the determination of amino acids and fatty acids in rat tissues. Thus a sample of rat fat was saponified yielding 14·641 g fatty acids and to this were added 0·2163 g palmitic acid (q_1) enriched to 21·5 atoms per cent with deuterium ($a_1(0)$). Palmitic acid was separated and purified, and its deuterium content found to be 1·28 atoms per cent ($a_1(t)$). Therefore the amount of palmitic acid in the sample (using equation 3.2)

is
$$Q_1 = 0·2163 \left(\frac{21·5}{1·28} - 1 \right) g$$
$$= 3·42 \text{ g}$$

Therefore, the fraction of palmitic acid in the sample

$$= \frac{3 \cdot 42 \times 100}{14 \cdot 641} \% = 23 \cdot 4\%$$

The above is the treatment of the simple case for one isotope and one substance. More complex situations are discussed by Gest, Kamen, and Reiner (1947). The main value of this technique is in the field of chemical analysis, as in the above example, but the concepts are implied in the determination of compartmental sizes of theoretical models.

3.4 One compartment system

A one compartment system is the simplest system to study. The theoretical model is shown in Figure 3.2. The differential equation

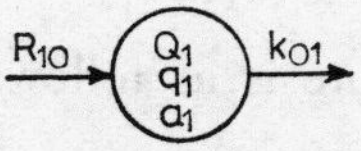

Fig. 3.2 A one compartment system: theoretical model.

describing the behaviour of Q_1 is

$$\frac{d(Q_1)}{dt} = R_{10} - k_{01} Q_1 \tag{3.4}$$

Because the system is in a steady state, then

$$\frac{d(Q_1)}{dt} = 0 \ , \quad \text{so that} \quad R_{10} = k_{01} Q_1$$

i.e. the input and output rates are equal. If tracer is injected into the compartment at $t = 0$, then its behaviour is given by

$$\frac{d(q_1)}{dt} = -k_{01} q_1 \tag{3.5}$$

Again, as in the previous section, $d(q_1)/dt$ can be replaced by $d(a_1)/dt \cdot Q_1$, and q_1 by $a_1 Q_1$ so that equation 3.5 becomes

$$\frac{d(a_1)}{dt} \cdot Q_1 = -k_{01} a_1 Q_1$$

or
$$\frac{d(a_1)}{dt} = -k_{01}a_1 \qquad (3.6)$$

Rearrangement and integration of equation 3.6 between $a_1 = a_1(0)$ when $t = 0$ and $a_1 = a_1$ when $t = t$ gives

$$-k_{01}\int_0^t dt = \int_{a_1(0)}^{a_1} \frac{d(a_1)}{a_1}$$

$$-k_{01}t = \ln a_1 - \ln a_1(0) = \ln\left(\frac{a_1}{a_1(0)}\right)$$

where ln represents logarithms to base e.

Hence,
$$\frac{a_1}{a_1(0)} = e^{-k_{01}t} \qquad (3.7)$$

or,
$$a_1 = a_1(0) . e^{-k_{01}t} \qquad (3.8)$$

An alternative procedure is as follows. Equation 3.6 can be rearranged to give

$$\frac{d(a_1)}{a_1} = -k_{01} . dt$$

which on integration produces

$$\ln a_1 = -k_{01}t + K$$

K is a constant, whose value depends on the initial conditions. When $t = 0$, $a_1 = a_1(0)$, so that

$$K = \ln a_1(0)$$

Hence
$$\ln \frac{a_1}{a_1(0)} = -k_{01}t$$

$$\frac{a_1}{a_1(0)} = e^{-k_{01}t}$$

or,
$$a_1 = a_1(0) . e^{-k_{01}t} \quad \text{(i.e. equation 3.8)}$$

Thus if the specific activity a_1 is determined and the data plotted with respect to time, an exponential curve will result (Fig. 3.3).

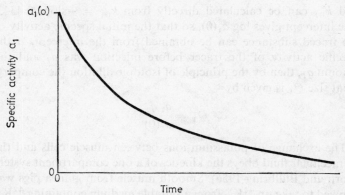

Fig. 3.3 A one compartment system. The exponential fall in concentration of tracer added to the system.

It is often convenient to express equation 3.8 in terms of $\log_{10}$, so that it becomes

$$\log a_1 = \log a_1(0) - 0.4343\, k_{01}\, t$$

If use is made of this equation and the data is plotted on semi-logarithmic graph paper, a straight line will result as in Figure 3.4. The slope of this straight line is given by

$$-\frac{(\log a_1(t_1) - \log a_1(t_2))}{(t_2 - t_1)}$$

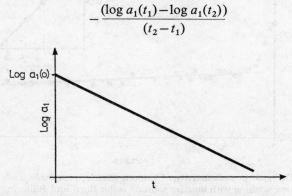

Fig. 3.4 A one compartment system. The straight line which results by plotting the data of Figure 3.3 on semi-log graph paper.

and k_{01} can be calculated directly from $k_{01} = -\text{slope}/0.4343$. The intercept gives log $a_1(0)$, so that the initial specific activity of the traced substance can be obtained from the intercept. If the specific activity of the tracer before injection was a_T and the amount q_T, then by the principle of isotopic dilution the compartment size Q_1 is given by

$$Q_1 = \frac{a_T q_T}{a_1(0)}$$

The exchange of potassium ions between muscle cells and the extracellular fluid obeys the kinetics of a one compartment system (Born and Büllbring 1956). Smooth muscle from guinea pigs was allowed to take up $^{42}K^+$ from a suitable medium containing $^{42}K^+$ until the radioactivity in the muscle had become constant. They were transferred to a constantly flowing non-radioactive medium. Apart from the first two minutes (probably due to washout from extracellular space) the subsequent loss of $^{42}K^+$ over five hours follows a single exponential curve (Fig. 3.5). The rate constant for

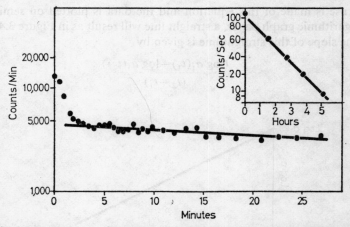

Fig. 3.5 Decrease in radioactivity of smooth muscle, previously loaded with $^{42}K^+$, during washing with inactive solution (after Born and Büllbring 1956). (Reproduced with the kind permission of the authors and the Editorial Board of the *Journal of Physiology*.)

potassium exchange was calculated from the slope of this line and its value was 1·14 hr.$^{-1}$. A second example of this type of system is given in Section 3.6 where it is used to illustrate another method for analyzing data which fits a single exponential term.

3.5 Simple two compartment system

If tracer is injected into compartment 1 of Figure 3.6 then the mathematical equations for this system in terms of total activities

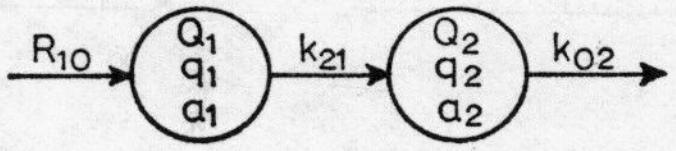

Fig. 3.6 A simple two compartment system: theoretical model.

are

$$\frac{d(q_1)}{dt} = -k_{21}q_1$$

$$\frac{d(q_2)}{dt} = k_{21}q_1 - k_{02}q_2$$

and in terms of specific activities

$$\frac{d(a_1)}{dt} = -k_{21}a_1$$

$$\frac{d(a_2)}{dt} = k_{21}\frac{Q_1}{Q_2}a_1 - k_{02}a_2$$

The solution of these equations is set out in Section A2 using the Laplace transform. The results are

$$a_1 = a_1(0).e^{-k_{21}t}$$

$$a_2 = \frac{k_{21}a_1(0)Q_1}{(k_{21}-k_{02})Q_2}(e^{-k_{02}t}-e^{-k_{21}t})$$

or more concisely

$$a_1 = X_1 e^{-\lambda_1 t}$$

$$a_2 = X_2(e^{-\lambda_2 t}-e^{-\lambda_1 t})$$

where $X_1 = a_1(0)$, $\lambda_1 = k_{21}$, $X_2 = [k_{21}a_1(0) Q_1/(k_{21}-k_{02})Q_2]$, and $\lambda_2 = k_{02}$. The plot of specific activities in both compartments is shown in Figure 3.7 for different values of k_{21} and k_{02}.

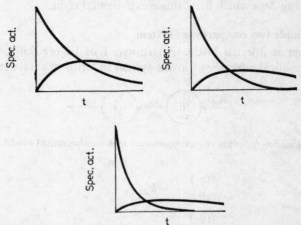

Fig. 3.7 The two compartment system of Figure 3.6. Specific activities of tracer in the two compartments showing how different values of the rate constants affect the shapes of the curves.

The procedure normally used for the estimation of the four parameters Q_1, Q_2, k_{02}, and k_{21} from these curves is as follows. a_1 can be treated as a single exponential in the manner of Section 3.4 to obtain $a_1(0)$, and hence Q_1 and k_{21}. The curve for a_2 is analyzed by one of the methods of Chapter 6 allowing k_{02}, k_{21}, and $k_{21} a_1(0) Q_1/(k_{21}-k_{02}) Q_2$ to be calculated. Thus all the parameters of the system can be determined.

3.6 One compartment system—uptake from a constant source

The theoretical model is shown in Figure 3.8 and in this case the specific activity a_0 of the first compartment remains constant. The rate of flow of unlabelled substance through compartment 1 is given by

$$\frac{d(Q_1)}{dt} = R_{10}-k_{01} Q_1$$

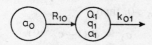

Fig. 3.8 A one compartment system in which tracer is added to compartment 1 at a constant rate R_{10}.

and since for the steady state $d(Q_1)/dt = 0$

then
$$R_{10} = k_{01} Q_1 \qquad (3.9)$$

The flow of tracer is described by

$$\frac{d(Q_1)}{dt} = a_0 R_{10} - a_1 k_{01} Q_1 \qquad (3.10)$$

Substituting equations 3.1 and 3.9 into equation 3.10

$$\frac{d(a_1)}{dt} = k_{01}(a_0 - a_1)$$

Integration by one of the methods of Section A1 or A2 yields

$$a_1 = a_0(1 - e^{-k_{01}t}) \qquad (3.11)$$

Alternatively, we can consider the system as a special case of the two compartment system (Figs. 3.6 and 3.9), in which compartment 0 approaches an infinite size. The equation describing the kinetics of tracer in compartment 1 is

$$a_1 = \frac{k_{10} Q_0 a_0}{(k_{10} - k_{01}) Q_1} (e^{-k_{01}t} - e^{-k_{10}t})$$

Since compartment 1 is in a steady state, then

$$k_{10} Q_0 = k_{01} Q_1$$

Therefore,
$$a_1 = \frac{k_{01} a_0}{(k_{01} - k_{01})} (e^{-k_{01}t} - e^{-k_{10}t})$$

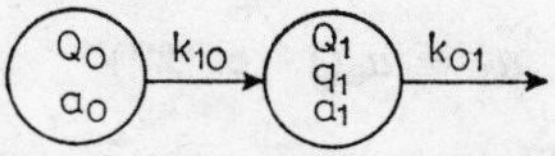

Fig. 3.9 Uptake from a constant source as a special case of a two compartment system. Tracer in compartment 0 remains at a constant specific activity.

Now as $Q_0 \to \infty$, $k_{10} \to 0$, then

$$a_1 = a_0(1 - e^{-k_{01}t}) \qquad (3.12)$$

which is identical to equation 3.11.

Equation 3.11 or 3.12 can be rearranged in two steps,

$$\frac{a_1}{a_0} = (1 - e^{-k_{01}t})$$

$$\left(1 - \frac{a_1}{a_0}\right) = e^{-k_{01}t}$$

a_0 is the value reached by a_1 when $t \to \infty$ i.e. $a_0 = a_1(\infty)$. The latter equation is similar in form to equation 3.7, and if the data (a_1) are plotted as $(1 - a_1/a_1(\infty))$ against t on semi-logarithmic graph paper a straight line will result with intercept of value 1 and slope equal to $-0\cdot4343\, k_{01}$.

London and Rittenberg (1950) used this method to determine the rate of cholesterol biosynthesis in man. They labelled the body water by means of a large oral dose of deuterium oxide and maintained the deuterium at a constant concentration for a month by giving smaller daily doses. Cholesterol was synthesized from precursors which had taken up deuterium from the body water. Hence, they could isolate serum cholesterol at suitable times and measure its deuterium content. The results, plotted as indicated above, are reproduced in Figure 3.10, from which the half-life for cholesterol was calculated to be about eight days.

The previous procedure is often unreliable, especially when it is difficult to know that one has approximated adequately to infinity. Bleehan and Fisher (1954) have published a better procedure which can be applied to any expression containing a single exponential provided experimental values are available at equal intervals of time. Thus we can take equation 3.11

$$a_1 = a_0(1 - e^{-k_{01}t})$$

and write it in the form

$$y = p + q \cdot e^{-kt}$$

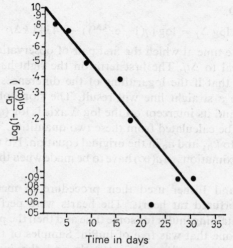

Fig. 3.10 Deuterium concentration in serum cholesterol digitonide plotted semi-logarithmically (after London and Rittenberg 1950). (Reproduced with the kind permission of the authors and the *Journal of Biological Chemistry*.)

Let $y_1, y_2, \ldots y_n$ be the experimental values of y measured at equal time intervals, Δt. We can then calculate the series of differences

$$\Delta_1 = y_1 - y_2$$

$$\Delta_2 = y_2 - y_3$$

$$\cdots\cdots\cdots\cdots$$

$$\Delta_{n-1} = y_{n-1} - y_n$$

In general,

$$\Delta_j = y_j - y_{j+1}$$
$$= q(e^{-kj\Delta t} - e^{-k(j+1)\Delta t})$$
$$= q(1 - e^{-k\Delta t}) \cdot e^{-k\Delta tj}$$

Taking logarithms to base e,

$$\ln\Delta_j = \ln(q(1 - e^{-k\Delta t})) - k\Delta tj$$

or to base 10,

$$\log \Delta_j = \log (q(1 - e^{-k\Delta t})) - 0 \cdot 4343\ k\Delta tj$$

Note that the time at which the first pair of observations is made, t_j, is identical to Δtj. The first term on the right-hand side is a constant so that if the logarithms of the differences are plotted against time a straight line will result. The slope of this line is $-0 \cdot 4343\ k$ and its intercept on the log Δ axis is $\log (q(1 - e^{-k\Delta t}))$. k and q can be calculated from these two quantities. These values correspond to k_{01} and a_0 in the original equation. No assumptions about approximations to $a(\infty)$ have to be made when this procedure is used.

Bleehan and Fisher used their procedure to measure inulin spaces in perfused rat hearts. The hearts were perfused with a medium containing inulin for some time and then the perfusate was changed to one that was free of inulin. Samples of the perfusate were collected over 10 min. intervals and the inulin content measured. The model of the system is the simple case of a one compartment system, and application of the above theory produces the same result:

$$\log \Delta_j = \log (q(1 - e^{-k\Delta t})) - 0 \cdot 4343\ kt_j$$

The results were then plotted as shown in Figure 3.11. Each point

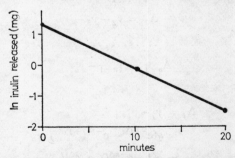

Fig. 3.11 The semi-logarithmic relation between inulin appearance in the medium after transfer of the perfused heart from inulin-containing to inulin-free medium and the time of commencement of successive periods of 10 min. (from Bleehan and Fisher 1954). (Reproduced with the kind permission of the authors and the Editorial Board of the *Journal of Physiology*.)

is the mean of five experiments and although there are only three points they fit a straight line very well. From the slope of the line the rate constant for the loss of inulin was calculated; $k = 0.14$ min.$^{-1}$. The amount of inulin initially present was calculated from the intercept (4.75 mg), and knowing the inulin concentration of the first perfusate (25 mg/ml), the inulin space was calculated (0.19 ml). Since the mean weight of the hearts was 0.68 g, the inulin space represents 28% v/w. This value is approximately the same as other accepted values for the extracellular space in tissues.

3.7 Constant infusion

This technique has occasionally been used in order to label compartments to a high specific activity, but the specific activity-time curves do not appear to have ever been analyzed to determine parameters. The system is shown in Figure 3.12. Assuming that

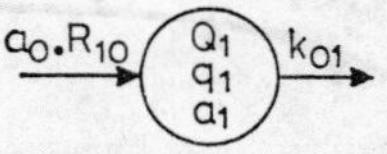

Fig. 3.12 A one compartment system in which tracer is infused at a constant rate $a_0 . R_{10}$.

the influx of tracer does not disturb the value of Q_1, then the behaviour of tracer is described by

$$\frac{d(q_1)}{dt} = R_{10} a_0 - q_1 k_{01}$$

which on integration yields

$$q_1 = \frac{R_{10} a_0}{k_{01}} (1 - e^{-k_{01}t})$$

but since $R_{10} = Q_1 k_{01}$ and $q_1 = a_1 Q_1$ then

$$a_1 = a_0 (1 - e^{-k_{01}t})$$

Constant infusion of tracer was probably first used by Hevesey and Hahn (1940) in order to obtain high concentrations of ^{32}P-phosphate in plasma (Fig. 3.13). Hevesey and Hahn were not interested in analyzing these data, but the author has taken their

D

results and fitted them, using a digital computer program, to the equation

$$a_1 = a_0(1 - e^{-k_{01}t})$$

The results are

$$a_0 = 0.142 \pm 0.0015 \quad \text{specific activity units}$$
$$k_{01} = 0.0352 \pm 0.0020 \quad \text{min.}^{-1}$$

Hence we can now estimate that the rate constant for the turnover of inorganic phosphate in rabbit plasma during the experiment of Hevesey and Hahn was 0.0352 min.$^{-1}$, or alternatively, the half-life was 19.5 min.

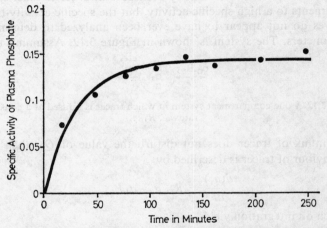

Fig. 3.13 Specific activity of plasma phosphate during the infusion of ^{32}P-phosphate at a constant rate. ● data of Hevesey and Hahn (1940), — curve fitted by the present author.

3.8 Multiple dosage

In the model of Figure 3.14 it is assumed that input of tracer does not disturb the value of Q_1. The amount of tracer in the system is described by

$$\frac{d(q_1)}{dt} = -k_{01}q_1$$

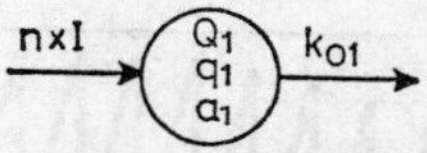

Fig. 3.14 A one compartment system in which tracer is added in n equal doses at equal time intervals.

An amount of tracer is injected at $t = 0$ so that the specific activity is $a_1(0)$, and at a time t_1 the specific activity in the compartment is

$$a_1(t_1) = a_1(0).e^{-k_{01}t}$$

At this time a second injection is given so that the specific activity at a very small time (δt) afterwards becomes

$$a_1(t_1+\delta t) = a_1(0)(1+e^{-k_{01}t})$$

and the value after a second time interval is

$$a_1(t_2) = a_1(0)(1+e^{-k_{01}t}).e^{-k_{01}}$$

A third injection gives a specific activity of

$$a_1(t_2+\delta t) = a_1(0)(1+e^{-k_{01}t}).e^{-k_{01}t}+a_1(0)$$
$$= a_1(0)(1+e^{-k_{01}t}+e^{-2k_{01}t})$$

and after the nth injection it would become

$$a_1(t_n) = a_1(0)(1+e^{-k_{01}t}+e^{-2k_{01}t}+\cdots+e^{-(n-1)k_{01}t})$$
$$= \frac{a_1(0)(1-e^{-nk_{01}t})}{(1-e^{-k_{01}t})} \tag{3.13}$$

As $n\to\infty$, $a_1(t_\infty)$ becomes $a_1(\infty)$, so that

$$a_1(\infty) = \frac{a_1(0)}{(1-e^{-k_{01}t})}$$

The complete sequence of events would appear as in Figure 3.15.

No formal analysis of such a system has been attempted, but Atkins et al. (1965) have used equation 3.13 to correct for initial enrichment $a_1(0)$ after multiple dosage with ^{13}C-ascorbic acid. In this experiment it was necessary to label the ascorbic acid within the subject to a high value so that the excretion of the label (as

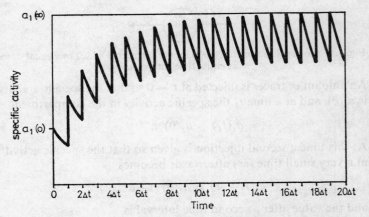

Fig. 3.15 Change of specific activity of tracer within a one compartment system while tracer is being added in equal doses at time intervals Δt. After an infinite number of doses the specific activity in the compartment oscillates to an equilibrium value $a_1(\infty)$.

^{13}C-ascorbic acid and ^{13}C-oxalic acid) could be followed for many weeks. A single dose of the tracer available of such a value that it would not upset the steady state (0·71 mg/kg body weight/24 hr., 48 atoms % excess) would not have produced the required enrichment of ^{13}C. Hence it was necessary to give the labelled ascorbic acid in 32 doses at 6-hourly intervals. It was then necessary to correct for the label which had been excreted during the eight days.

3.9 Two compartment closed system

The theoretical model for this system is shown in Figure 3.16. The differential equations for the mother substance, taking each compartment in turn, are

$$\frac{d(Q_1)}{dt} = k_{12} Q_2 - k_{21} Q_1 = 0$$

$$\frac{d(Q_2)}{dt} = k_{21} Q_1 - k_{12} Q_2 = 0$$

If tracer is present in both compartments then the equations for the tracer are

$$\frac{d(q_1)}{dt} = k_{12}q_2 - k_{21}q_1$$

$$\frac{d(q_2)}{dt} = k_{21}q_1 - k_{12}q_2$$

Using equation 3.1

$$\frac{d(a_1)}{dt} = k_{12}a_2\frac{Q_2}{Q_1} - k_{21}a_1$$

$$\frac{d(a_2)}{dt} = k_{21}a_1\frac{Q_1}{Q_2} - k_{12}a_2$$

Note that $a_1 Q_1 + a_2 Q_2$ is constant.

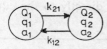

Fig. 3.16 A two compartment closed system: theoretical model. Material is exchanged between the two compartments and does not leave the system.

If tracer is injected initially into compartment 1 only, then at $t = 0$, $a_1 = a_1(0)$ and $a_2 = 0$. Solution of the equations is more difficult than for previous systems but can be achieved by any of the standard processes. The solution is obtained by using differential operators in Section A1. The result is

$$a_1 = \frac{k_{12}a_1(0)}{(k_{21}+k_{12})} + \frac{k_{21}a_1(0)}{(k_{21}+k_{12})} \cdot e^{-(k_{12}+k_{21})t}$$

$$a_2 = \frac{k_{21}a_1(0)Q_1}{(k_{21}+k_{12})Q_2}(1-e^{-(k_{12}+k_{21})t})$$

or more concisely

$$\left.\begin{array}{l} a_1 = X_0 + X_1e^{-\lambda t} \\ a_2 = X_2(1-e^{-\lambda t}) \end{array}\right\} \qquad (3.14)$$

where
$$X_0 = \frac{k_{12}\,a_1(0)}{(k_{21}+k_{12})}$$

$$X_1 = \frac{k_{21}\,a_1(0)}{(k_{21}+k_{12})}$$

$$X_2 = \frac{k_{21}\,a_1(0)\,Q_1}{(k_{21}+k_{12})\,Q_2}$$

and
$$\lambda = k_{12}+k_{21}$$

A plot of specific activities for the two compartments is shown in Figure 3.17. There are several methods for obtaining estimates of the four parameters for this system—Q_1, Q_2, k_{12}, and k_{21}. Two are given below, and a third procedure (Bleehan and Fisher 1954) has been presented in section 3.6.

Curve analysis. By the methods of Chapter 6 it is possible to fit the data of Figure 3.17 directly to the equations 3.14 and thus

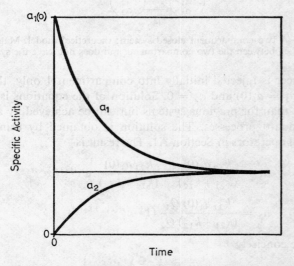

Fig. 3.17 A two compartment closed system. Specific activities within the two compartments after tracer is added to compartment 1 at zero time. The two specific activities reach the same value after a long time period.

obtain values for X_0, X_1, X_2, and λ. If the initial specific activities and amount of tracer are a_T and q_T respectively, then

$$Q_1 = \frac{a_T q_T}{a_1(0)}$$

and thus the four parameters can be estimated.

Prentice, Siri, and Joiner (1952) used the model of Figure 3.16 to represent the exchange of tritium-labelled water between plasma and ascitic fluid. Tritium oxide was injected into the ascitic fluid of a patient suffering from portal cirrhosis, and the specific activities of plasma and ascitic fluid were determined. The results are shown in Figure 3.18. Graphical analysis, as described in Chapter 6, and then calculation of the model parameters gave the values: $Q_1 = 44 \cdot 5$ l, $Q_2 = 6 \cdot 6$ l, $k_{21} = 0 \cdot 40$ hr.$^{-1}$, $k_{12} = 0 \cdot 06$ hr.$^{-1}$. Q_2 ($6 \cdot 6$ l) represents the volume of ascitic fluid and Q_1 ($44 \cdot 5$ l) represents total body fluid minus ascitic fluid. The weight of the patient was

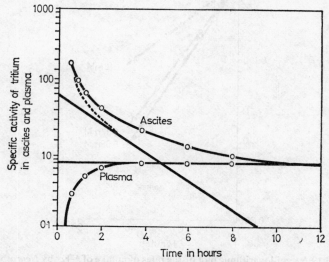

Fig. 3.18 Semi-logarithmic plot showing the transfer of tritiated water from ascites into plasma (from Prentice, Siri, and Joiner 1952). (Reproduced with the kind permission of the authors and the *American Journal of Medicine*.)

77·3 kg i.e. Q_1 represents $44·5/(77·3-6·6)\times 100\%$ which is 63% of body weight. The usually accepted value for total body water is about 600 ml/kg. The turnover of ascitic fluid was $(44·5\times 0·06)$ or $(0·40\times 6·6)$ l/hr. i.e. 2·65 l/hr.

Rearrangement of the equations. When $t = \infty$, $e^{-\lambda t} = 0$, then $a_1(\infty) = X_0$ and $a_2(\infty) = X_2$. The relative specific activity

$$\frac{a_1}{a_1(\infty)} = \frac{X_0 + X_1 e^{-\lambda t}}{X_0} = 1 + \frac{X_1 e^{-\lambda t}}{X_0}$$

Rearranging,

$$\left(\frac{a_1}{a_1(\infty)} - 1\right) = \frac{X_1 e^{-\lambda t}}{X_0}$$

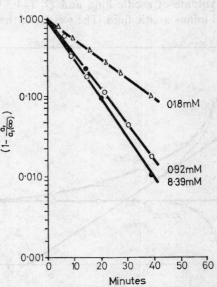

Fig. 3.19 A semi-logarithmic plot of the rates of uptake of $^{42}K^+$ by *Neurospora* measured at different external potassium ion concentrations (after Slayman and Tatum 1965). (Reproduced with kind permission of the authors and the publishers of *Biochimica et Biophysica Acta*.)

Taking logarithms,

$$\ln\left(\frac{a_1}{a_1(\infty)}-1\right) = \ln\frac{X_1}{X_0}-\lambda t$$

$$\log\left(\frac{a_1}{a_1(\infty)}-1\right) = \log\frac{X_1}{X_0}-0\cdot4343\,\lambda t$$

Hence, plotting $[(a_1/a_1(\infty))-1]$ on semi-logarithmic graph paper will give estimates of X_0, X_1, and λ. $a_1(0)$, $a_1(\infty)$, and $a_2(\infty)$ are also determined from Figure 3.17, so λ, X_0, X_1, and X_2 are known and the four parameters of the model can be estimated.

An example of this method is given by Slayman and Tatum (1965) who suspended *Neurospora* cells in a $^{42}K^+$ medium, then at suitable times removed samples of cells, washed them and measured their radioactivity. Some of their results are shown in Figure 3.19, the lines referring to different potassium concentrations. The data for each line are unfortunately not produced, only the calculated influx rates corresponding to chosen potassium ion

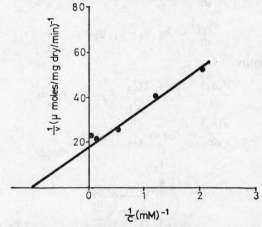

Fig. 3.20 A Lineweaver–Burk plot showing the linear relation between the reciprocal of the steady state influx rate and the reciprocal of the external potassium ion concentration (from Slayman and Tatum 1965). (Reproduced with the kind permission of the authors and the publishers of *Biochimica et Biophysica Acta*.)

concentrations are given. These authors then proceed to plot the reciprocal of influx rate against the reciprocal of potassium ion concentration (Lineweaver–Burk plot) to demonstrate that influx of potassium ions appears to follow Michaelis–Menton kinetics (Fig. 3.20).

Note that for all methods it is necessary to have data concerning both compartments in order to determine all the parameters of the system.

If tracer were injected into compartment 2, then because the system is symmetrical the treatment is exactly the same as above except that the subscripts 1 and 2 are interchanged. The equations and the estimation of the parameters are otherwise identical.

3.10 Two compartment open system

The differential equations for the tracer in the model of Figure 3.21, taking each compartment in turn, are

$$\frac{d(q_1)}{dt} = k_{12}q_2 - k_{21}q_1 - k_{01}q_1$$

$$\frac{d(q_2)}{dt} = k_{21}q_1 - k_{12}q_2$$

Using equation 3.1

$$\frac{d(a_1)}{dt} = k_{12}a_2\,\frac{Q_2}{Q_1} - k_{21}a_1 - k_{01}a_1$$

$$\frac{d(a_2)}{dt} = k_{21}a_1\,\frac{Q_1}{Q_2} - k_{12}a_2$$

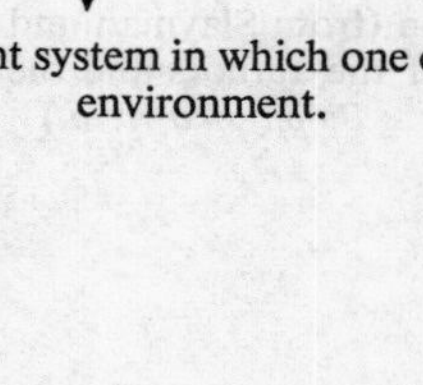

Fig. 3.21 A two compartment system in which one compartment is open to the environment.

The solution of these equations is performed in Section A2 for the case when tracer is injected into compartment 1, and the results are

$$a_1 = \frac{a_1(0)}{(\lambda_1 - \lambda_2)} \left[(\lambda_1 - k_{12}) e^{-\lambda_1 t} + (k_{12} - \lambda_2) e^{-\lambda_2 t} \right]$$

$$a_2 = \frac{a_1(0) k_{21} Q_1}{(\lambda_1 - \lambda_2) Q_2} \left[e^{-\lambda_1 t} - e^{-\lambda_2 t} \right]$$

where

$$\lambda_1 = \frac{-(k_{12} + k_{21} + k_{01}) - \sqrt{(k_{12} + k_{21} + k_{01})^2 - 4 k_{01} k_{12}}}{2}$$

$$\lambda_2 = \frac{-(k_{12} + k_{21} + k_{01}) + \sqrt{(k_{12} + k_{21} + k_{01})^2 - 4 k_{01} k_{12}}}{2}$$

It will be obvious that as the system becomes more complex the expressions for the exponential coefficients also increases in complexity. Concisely, the equations may be written in the form

$$a_1 = X_1 e^{-\lambda_1 t} + X_2 e^{-\lambda_2 t} \qquad (3.15)$$

$$a_2 = X_3 e^{-\lambda_1 t} - X_3 e^{-\lambda_2 t}$$

where

$$X_1 = \frac{a_1(0)(\lambda_1 - k_{12})}{(\lambda_1 - \lambda_2)}$$

$$X_2 = \frac{a_1(0)(k_{12} - \lambda_2)}{(\lambda_1 - \lambda_2)}$$

$$X_3 = \frac{a_1(0) k_{21} Q_1}{(\lambda_1 - \lambda_2) Q_2}$$

The specific activity–time curves are of the form shown in Figure 3.22. There are no simple means for determining the constants X_1, X_2, X_3, λ_1, and λ_2 from the data. One of the techniques of curve analysis presented in Chapter 6 must be used, and then the parameters of the model can be determined.

Injection of tracer into compartment 1 is the most common case for this system, but it is equally possible to obtain equations for

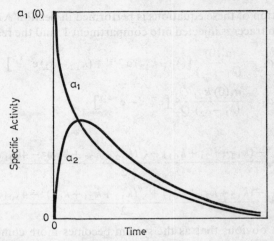

Fig. 3.22 Specific activities within the two compartments of the system of Figure 3.21 after tracer has been added to compartment 1 at zero time.

a_1 and a_2 when tracer is put into compartment 2 alone, or when it is put into both compartments.

Paterson and Harrison (1968) have used this system to explain the distribution of cortisol within the sheep, and have used the case where both compartments are initially labelled to the same specific activity. This was achieved by infusing 1,2-^{3}H-cortisol intravenously into the animal for four hours until the specific activity of the plasma cortisol had become constant. The resulting equations in this case are of the same form as before. For example, equation 3.15 remains as

$$a_1 = X_1 \, e^{-\lambda_1 t} + X_2 \, e^{-\lambda_2 t}$$

but now

$$X_1 = \frac{a_1(0)\,(k_{01} - \lambda_2)}{(\lambda_1 - \lambda_2)}$$

and

$$X_2 = \frac{a_1(0)\,(\lambda_1 - k_{01})}{(\lambda_1 - \lambda_2)}$$

After cessation of infusion plasma samples were taken over 60 min., and their specific activities were measured. The data corresponds to that for compartment 1 in the model of Figure 3.21 and could be fitted to their two exponential expression for a_1. Pooled data from eight experiments on one ewe are presented in Figure 3.23, and curve fitting gave

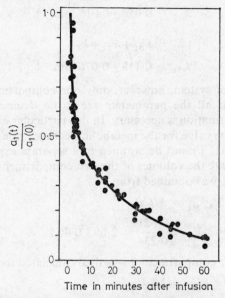

Fig. 3.23 Relative specific activity of ^{3}H-cortisol after cessation of infusion of labelled cortisol. Pooled data from eight experiments on one sheep (after Paterson and Harrison 1968). (Reproduced with the kind permission of the authors and the editor of the *Journal of Endocrinology*.)

$$\frac{X_1}{a_1(0)} = 0.568 \pm 0.046$$

$$\frac{X_2}{a_1(0)} = 0.432 \pm 0.042$$

$$\lambda_1 = 0.420 \pm 0.069 \text{ min.}^{-1}$$

$$\lambda_2 = 0.033 \pm 0.004 \text{ min.}^{-1}$$

From
$$\frac{X_1}{a_1(0)} = \frac{(k_{01} - \lambda_2)}{(\lambda_1 - \lambda_2)}$$

$$k_{01} = 0 \cdot 234 \pm 0 \cdot 04 \text{ min.}^{-1}$$

from
$$\lambda_1 \lambda_2 = k_{01} k_{12}$$

$$k_{12} = 0 \cdot 056 \pm 0 \cdot 016 \text{ min.}^{-1}$$

and from
$$\lambda_1 + \lambda_2 = k_{01} + k_{21} + k_{12}$$

$$k_{21} = 0 \cdot 118 \pm 0 \cdot 076 \text{ min.}^{-1}$$

For this type of system, however, only one compartment can be sampled and if all the parameters are to be determined some additional information is necessary. In this particular example, the authors used the value for the metabolic clearance rate (MCR) for cortisol ($= 0 \cdot 621$ l/min.) determined in a separate experiment in order to calculate the volumes of the two compartments. That for compartment 1 was obtained from

$$V_1 k_{01} = MCR$$

i.e.
$$V_1 = \frac{0 \cdot 621}{0 \cdot 234} = 2 \cdot 65 \pm 0 \cdot 46 \text{ l}$$

The volume for compartment 2 was then calculated from

$$k_{12} V_1 = k_{21} V_2$$

i.e.
$$V_2 = \frac{0 \cdot 118 \times 2 \cdot 65}{0 \cdot 056} = 5 \cdot 61 \pm 4 \cdot 0 \text{ l}$$

The mean plasma cortisol concentration was also determined ($= 22 \cdot 6$ μg/l), hence using V_1 and V_2 the two compartment sizes were found to be

$$Q_1 = 60 \pm 10 \ \mu\text{g}$$

and
$$Q_2 = 127 \pm 91 \ \mu\text{g}$$

The volume of compartment 1 corresponds approximately to that for whole blood in the sheep, but the volume of compartment 2 appears to have no physiological interpretation.

3.11 Three compartment system

The system of Figure 3.24 was one of the first to be treated analytic-

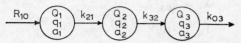

Fig. 3.24 The simplest three compartment system. Material in each compartment is only received from a previous one and only delivered to the next one in series.

ally (Artom, Sarzana, and Segre 1938). The differential equations for the concentration of tracer in this system are

$$\frac{d(q_1)}{dt} = -k_{21} q_1$$

$$\frac{d(q_2)}{dt} = k_{21} q_1 - k_{32} q_2$$

$$\frac{d(q_3)}{dt} = k_{32} q_2 - k_{03} q_3$$

and for the concentration of tracer

$$\frac{d(a_1)}{dt} = -k_{21} a_1$$

$$\frac{d(a_2)}{dt} = k_{21} \frac{Q_1}{Q_2} a_1 - k_{32} a_2$$

$$\frac{d(a_3)}{dt} = k_{32} \frac{Q_2}{Q_3} a_2 - k_{03} a_3$$

The solution to this set of equations is given in Section A1 and is as follows

$$a_1 = X_1 e^{-\lambda_1 t}$$

$$a_2 = X_2 e^{-\lambda_1 t} + X_3 e^{-\lambda_2 t}$$

$$a_3 = X_4 e^{-\lambda_1 t} + X_5 e^{-\lambda_2 t} + X_6 e^{-\lambda_3 t}$$

where
$$\lambda_1 = k_{21}$$
$$\lambda_2 = k_{32}$$
$$\lambda_3 = k_{03}$$
$$X_1 = a_1(0)$$
$$X_2 = \frac{k_{21}\, a_1(0)\, Q_1}{(k_{32}-k_{21})\, Q_2}$$
$$X_3 = \frac{k_{21}\, a_1(0)\, Q_1}{(k_{21}-k_{32})\, Q_2}$$
$$X_4 = \frac{k_{32}\, k_{21}\, a_1(0)\, Q_1}{(k_{03}-k_{32})\,(k_{32}-k_{21})\, Q_3}$$
$$X_5 = \frac{k_{32}\, k_{21}\, a_1(0)\, Q_1}{(k_{21}-k_{03})\,(k_{32}-k_{21})\, Q_3}$$
$$X_6 = \frac{k_{32}\, k_{21}\, a_1(0)\, Q_1}{(k_{32}-k_{03})\,(k_{21}-k_{03})\, Q_3}$$

Figure 3.25 shows the specific activity–time curves for this type of model.

Curve analysis of the data obtained in all three compartments will yield nine constants, together with a value for the amount of tracer added to the system. There are six parameters to be calculated (k_{21}, k_{32}, k_{03}, Q_1, Q_2, and Q_3) and therefore six pieces of information are sufficient. Four out of the available ten experimental values are therefore unnecessary, and the data is said to be degenerate. From the three relationships

$$\frac{X_4}{X_5} = \frac{(k_{21}-k_{03})}{(k_{03}-k_{32})}$$

$$\frac{X_4}{X_6} = \frac{(k_{03}-k_{21})}{(k_{32}-k_{21})}$$

and
$$\frac{X_5}{X_6} = \frac{(k_{32}-k_{03})}{(k_{32}-k_{21})}$$

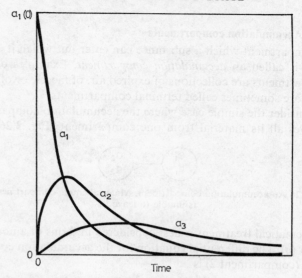

Fig. 3.25 Specific activities in the three compartments of the model of Figure 3.24 after tracer has been added to compartment 1 at zero time.

three simultaneous equations in k_{03}, k_{21}, and k_{32} can be obtained. Solution of these equations will provide values for the three rate constants so that the experimental values λ_1, λ_2, and λ_3 need not be used. Q_1 can be calculated using equation 3.3 i.e.

$$Q_1 = \frac{\text{(amount of tracer added)}}{a_1(0)}$$

$$= \frac{\text{(amount of tracer added)}}{X_1}$$

Q_2 can then be obtained from either X_2 or X_3, but only one of these quantities need be known. Likewise, Q_3 can be calculated from X_4 X_5, or X_6, all of which have already been used. Even though the data are degenerate, and part is not used, it is necessary to sample all three compartments to determine all the parameters.

E

3.12 Accumulation compartments

A compartment which a substance can enter but which it cannot leave is called an *accumulation compartment*. Examples of such compartments are collections of expired air, of faeces, or of urine. They are sometimes called terminal compartments.

Consider the simple case where the accumulation compartment receives all its material from one compartment (Fig. 3.26). The

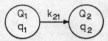

Fig. 3.26 An accumulation compartment. Material enters compartment 2 but is unable to leave it.

mathematical treatment is best considered in terms of amounts of tracer (q). The differential equation for the accumulation compartment (compartment 2) is

$$\frac{d(q_2)}{dt} = q_1 k_{21}$$

When $t = 0$, $q_1 = q_1(0)$. At all times $q_1 + q_2 = q_1(0)$ i.e. $q_1 = q_1(0) - q_2$.

Therefore, $$\frac{d(q_2)}{dt} = q_1(0) k_{21} - q_2 k_{21}$$

Solution of this equation by one of the methods of Section A1 or A2 yields

$$q_2 = q_1(0)(1 - e^{-k_{21}t})$$

An example of a more complex system, but one which includes accumulation compartments, is that of Cummings, King, and Martin (1967) in which the metabolism and excretion of paracetamol was studied. They measured the elimination of some of the metabolites of paracetamol in man after a dose of the drug. It is not stated in the original paper, but it is assumed that the paracetamol was given orally, and that the absorption occurs at a much greater rate than any of the subsequent transformations. If we consider only the formation and excretion of paracetamol sulphate, the model of this system would be as in Figure 3.27. The math-

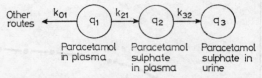

Fig. 3.27 A three compartment model, that contains an accumulation compartment, to determine some of the kinetics of paracetamol and paracetamol sulphate metabolism in man.

ematical model for this system is

$$\frac{d(q_1)}{dt} = -q_1(k_{01}+k_{21})$$

$$\frac{d(q_2)}{dt} = q_1 k_{21} - q_2 k_{32}$$

$$\frac{d(q_3)}{dt} = q_2 k_{32}$$

Solution for q_3 using one of the methods of Section A1 or A2 gives

$$q_3 = \frac{k_{21}q_1(0)}{(k_{01}+k_{21})} - \frac{k_{21}q_1(0)}{(k_{01}+k_{21}-k_{32})} e^{-k_{32}t} + \frac{k_{32}k_{21}q_1(0)}{(k_{01}+k_{21})(k_{01}+k_{21}-k_{32})} e^{-(k_{01}+k_{21})t}$$

After the subject had taken the paracetamol, consecutive urine samples were collected at 1·5-hr. intervals and the paracetamol sulphate content determined. The present author does not favour the mathematical treatment used in the original paper, but prefers to fit the data either to the sum of three exponentials, or directly to the three differential equations by using a suitable digital computer program. Both these methods are discussed in Chapter 6. The experimental data obtained on one of their subjects and the curve fitted by the second of the two procedures are shown in Figure 3.28.

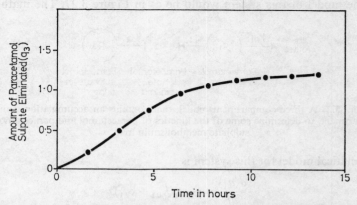

Fig. 3.28 Amount of paracetamol sulphate collected in compartment 3 of Figure 3.27. ● data of Cummings, King, and Martin (1967). — curve fitted by the present author.

The calculated parameters are:
rate constant for synthesis of paracetamol sulphate

$$= k_{21} = 0·090 \pm 0·0010 \text{ hr.}^{-1}$$

rate constant for elimination of paracetamol sulphate

$$= k_{32} = 0·753 \pm 0·0067 \text{ hr.}^{-1}$$

combined rate constant for the other methods of removal of paracetamol

$$= k_{01} = 0·257 \pm 0·0036 \text{ hr.}^{-1}$$

One point which must be noted, is that sampling of compartments must be made at comparable times. This is especially important when dealing with accumulation compartments. For example, it takes some time for urine to travel from the glomerulus in the kidney to the outside of the body.

3.13 Precursor–product relationship

The precursor–product relationship was first discussed in the classic paper of Zilversmit, Entenman, and Fishler (1943). The following treatment is taken from that paper, and the model for the system

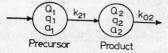

Precursor   Product

Fig. 3.29 A model to explain the precursor-product relationship. Material in compartment 2 is derived solely from that of compartment 1.

is shown in Figure 3.29. The specific activity in the first compartment is represented as a function of time $f(t)$ i.e. a_1. The equations describing the amount of tracer in this system are

$$\frac{d(q_1)}{dt} = -k_{21} a_1 Q_1$$

$$\frac{d(q_2)}{dt} = k_{21} a_1 Q_1 - q_2 k_{02} \qquad (3.16)$$

Because the various compartments are in a steady state

$$Q_1 k_{21} = Q_2 k_{02}$$

so that

$$k_{02} = \frac{Q_1 k_{21}}{Q_2} \qquad (3.17)$$

Substituting equation 3.17 in equation 3.16 and converting from total activities

$$Q_2 \frac{d(a_2)}{dt} = k_{21} a_1 Q_1 - a_2 Q_1 k_{21}$$

Rearranging,

$$\frac{d(a_2)}{dt} = \frac{k_{21} Q_1}{Q_2} (a_1 - a_2)$$

Using this relationship, three criteria can be defined if the material in compartment 2 is to be obtained solely from compartment 1:

(1) If $d(a_2)/dt$ is positive i.e. before a_2 reaches its maximum, then a_1 is greater than a_2.
(2) If $d(a_2)/dt = 0$, then $a_1 = a_2$. This means that the specific activity–time curves for the two components cross when the specific activity of the second reaches its maximum.
(3) If $d(a_2)/dt$ is negative i.e. after a_2 reaches its maximum, then a_2 is now greater than a_1.

An example of precursor–product relationships is provided by Hallberg (1965). He injected intravenously an emulsion of triglycerides (exogenous triglyceride) and then took samples of plasma at suitable time intervals. The concentrations of exogenous triglyceride, endogenous triglyceride, and endogenous lipoprotein were determined in each of these samples. Provided the initial injection of triglycerides was below a certain value (otherwise the kinetics were not linear), results were similar to those shown in Figure 3.30. It is apparent that the curve for each product crosses its predecessor at the maximum, and the results are consistent

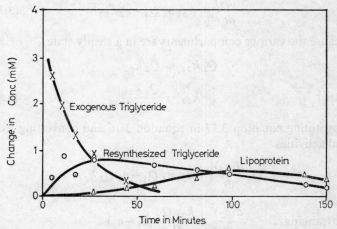

Fig. 3.30 Two precursor-product relationships showing the resynthesis of triglyceride from exogenous triglyceride, and lipoprotein synthesis from endogenous triglyceride (after Hallberg 1965; Figure 1B, data replotted on normal coordinates).

with the metabolic sequence exogenous triglyceride→ resynthesized triglyceride → synthesis of lipoprotein.

The precursor–product relationship can be used to determine whether a tissue constituent is wholly derived from the blood. Dayton (1959) measured the specific activity of the cholesterol of chick aorta at different times after administration of labelled cholesterol. The relation between specific activity of cholesterol and time is shown for plasma and aorta in Figure 3.31. Since the

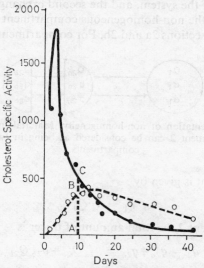

Fig. 3.31 Specific activities of thoracic aorta cholesterol and of plasma cholesterol (from Dayton 1959). (Reproduced with the kind permission of the editor of *Circulation Research*.)

maximum specific activity of aortic cholesterol was much less than the specific activity of the plasma at the same time, the cholesterol contributed by the plasma must have been diluted with unlabelled cholesterol of local origin. The ratio of the plasma contribution to the local contribution is $AB:BC$.

Other conditions that must apply if the precursor–product relationship is valid are (1) that the compartments must be homogeneous, and (2) that the true precursor must not be in rapid equilibrium with its own precursor.

3.14 Non-homogeneity

The concept of homogeneous compartments is fundamental to most of compartmental analysis. However, if a non-homogeneous compartment is suspected to be present, it can usually be represented by two sub-compartments. One of these exchanges rapidly with the rest of the system, and the second exchanges slowly with the first. Thus the non-homogeneous compartment in Figure 3.32 is divided into sections 2a and 2b. For compartment 1 the amount

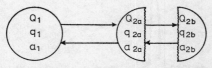

Fig. 3.32 Representation of non-homogeneity. Material in the non-homogeneous compartment 2 can be considered as being in two homogeneous compartments.

of tracer present is given by

$$q_1 = a_1 Q_1$$

For compartment 2 the total amount of tracer is

$$q_2 = q_{2a} + q_{2b} = a_{2a} Q_{2a} + a_{2b} Q_{2b}$$

If $\bar{a}_2$ is the mean specific activity of the substance in compartment 2, then

$$q_2 = \bar{a}_2 Q_2$$

Because Q_1 and Q_2 are exchanging rapidly, then a_{2a} is almost equal to a_1, so that

$$\bar{a}_2 Q_2 = a_1 Q_{2a} + a_{2b} Q_{2b}$$

Rearranging,

$$Q_{2b} a_{2b} = \bar{a}_2 Q_2 - a_1 Q_{2a}$$

If $\bar{a}_2$ and a_1 are expressed as functions of time, then

$$a_{2b}(t) = \bar{a}_2(t)\frac{Q_2}{Q_{2b}} - a_1(t)\frac{Q_{2a}}{Q_{2b}}$$

Now, if tracer is initially added to compartment 1, $a_{2b}(0)$ will be zero. During the experiment the specific activities in compartments 1 and 2 can be determined and plotted as functions. Extrapolation of these functions to $t = 0$ gives values for $\bar{a}_2(0)$ and $a_1(0)$. Thus, at $t = 0$

$$\bar{a}_2(0) Q_2 - a_1(0) Q_{2a} = 0$$

and

$$\frac{Q_{2a}}{Q_2} = \frac{\bar{a}_2(0)}{a_1(0)}$$

Hence, the sizes of the sub-compartments can be calculated. The above treatment is similar to one published by Wrenshall (1955).

In the few papers published so far that have used this procedure the non-homogeneous compartment has exchanged material with only one other compartment, and division of the compartment into two sub-compartments has been found to be adequate. Hart (1957) has however considered, theoretically, the extreme case where there is a concentration gradient within the non-homogeneous compartment. The compartment is then equivalent to an infinite number of sub-compartments.

Lax, Sidlofsky, and Wrenshall (1956) have used the above method in order to determine the rates of distribution of phosphorus in the rat between a central compartment (plasma) and fourteen peripheral ones (various tissues). The peripheral compartments were assumed to be non-homogeneous. They injected groups of rats with ^{32}P-labelled plasma and then at various times up to 410 min. each group was killed. The organs were divided into thirteen sections (heart, kidneys, brain, etc.), and the plasma phosphate divided into that which was precipitated by trichloracetic acid (approximately equivalent to lipoprotein) and that which was soluble (approximately equivalent to inorganic phosphate). This produced fourteen peripheral compartments and one central one, and for each of these ^{32}P and total P were determined. Specific activities were extrapolated to $t = 0$, so that $a_1(0)$, $\bar{a}_2(0)$, $\bar{a}_3(0)$, ... $\bar{a}_{15}(0)$ could be determined, and hence the values Q_{2a}, Q_{3a}, ... Q_{15a} were calculated. A curve was fitted to the data for the plasma inorganic phosphate and hence the total rate of transfer of P_i out of the central compartment was calculated. Using the values of $d(a_2)/dt$, $d(a_3)/dt$, etc. the individual transfer rates for each compartment were then obtained.

3.15 Time lags

In studying the metabolism of ^{131}I in which there is conversion to ^{131}I-thyroxine, or the metabolism of ^{59}Fe during which ^{59}Fe-haemoglobin may be synthesised, there will be a time interval between removal of the tracer from the plasma and its reappearance. Sharney *et al.* (1963) and Sharney *et al.* (1965) have discussed this situation and shown that the passage of material through the time delay can be represented as transport through an infinite number of vanishingly small compartments. This is a technical device and has no apparent physiological meaning.

Baker and Schotz (1964) and Schotz, Baker, and Charez (1964) injected 1-^{14}C-palmitate into the plasma of rats and then measured the rate at which labelled triglyceride appeared in the plasma. The triglyceride was synthesized within the liver from the free acid, and the authors observed a ten minute time interval between injection of the acid and first appearance of the labelled triglyceride. The experimental data could be fitted to a model system which included eight small compartments to represent the time lag; but the choice of eight was quite arbitrary and depended solely on the capacity of the computer used in analyzing the data. It is difficult to place a physiological interpretation on these eight small compartments.

A better technique is to incorporate the time delay directly into the model system. In a hydrodynamic model (discussed in Chapter 5) a delay between two compartments can be achieved if transport is by means of a glass coil. The material being transferred will take some time passing through the coil and the time delay will depend on the length of the coil. If analogue computers are used either for simulation (discussed in Chapter 5) or for curve fitting (discussed in Chapter 6) components can be used which will delay the electrical signals by whatever time interval is desired.

Curve fitting by digital computer programs may take two forms (discussed in Chapter 6). If data are fitted to sums of exponentials then the time delay can be incorporated into the function being fitted. Thus an exponential term would take the form

$$a = X \cdot e^{-\lambda(t - t_0)}.$$

where t_0 represents the time delay, and may or may not be one of the parameters of the model required to be estimated. An example of this procedure is presented in Section 4.2. On the other hand, data may be fitted to a mathematical model consisting of differential equations and these equations could then be integrated by numerical analysis during the curve fitting process. In such a case the values of the dependent variables $(a_1, a_2, \ldots)$ as they are calculated at each value of time (t_j) may be stored in an array or a matrix. Values at some previous time value $(t_j - t_0)$ can then be referred to when necessary. This technique has been used successfully by the present author on several occasions (unpublished work).

3.16 Lumping of compartments

If two compartments have the same, or very similar, rate constants k_{ij} then the changes in concentration of tracer during transfer between these compartments and a third compartment will be similar, and the two compartments can be treated together as a single compartment in the model system. Thus plasma triglycerides or plasma proteins can often be treated as one homogeneous compartment, whereas chemically they are not homogeneous, and it is, of course, very advantageous to be able to consider all the cells in a tissue (for example erythrocytes in plasma) as one compartment. Where it is a disadvantage e.g. if two tissues are exchanging with plasma and the compartments cannot be separated by injecting tracer into the plasma, it is occasionally possible to separate the compartments by placing the tracer into one of the other compartments. Sharney, Wasserman, and Gevirtz (1964) have produced a theoretical treatment discussing the conditions under which compartments may or may not be lumped.

3.17 Conclusions

Inspection of the systems discussed so far indicates some general conclusions. For example, all the integrated expressions for the concentration of tracer in an open system consist of a sum of exponential terms

$$a_i = \sum X_j e^{-\lambda_j t} \qquad (j = 1, 2, \ldots n)$$

If the system is closed the first term becomes a constant

$$a_i = X_1 + \sum X_j e^{-\lambda_j t} \qquad (j = 2, 3, \ldots n)$$

The number of terms in the expression is the same as the number of compartments in the system into which the tracer can enter.

For complete determination of all the parameters of a model it is necessary to have access to all compartments. If some information is missing e.g. if measurements are made on only one compartment, then only a partial solution of the system is possible and not all of the parameters can be determined. Usually this means that rate constants but not compartment sizes can be obtained.

Theoretically, the total number of all possible two compartment systems, including constant source and accumulation compartments but omitting duplicates due to symmetry, is twenty four. Sixteen of these theoretical models are given by Sharney *et al.* (1963). Thus, it is obvious that as the number of compartments, n, increases the number of alternative models becomes enormous. In practice, from physiological considerations, probably many of these models could not occur, but even so, a large number of models still remains. To obviate the unnecessary labour in solving each system from first principles it is sometimes possible to utilise the general solution for a general system of interconnected compartments. Such a solution is discussed in Section A4.

REFERENCES

ARTOM, C., SARZANA, G. and SEGRE, E. (1938). *Archs int. Physiol.*, **47**, 245. Influence des graisses alimentaires sur la formation des phospholipides dans les tissus animaux.

ATKINS, G. L., DEAN, B. M., GRIFFIN, W. J., SCOWEN, E. F. and WATTS, R. W. E. (1965). *Clin. Sci.*, **29**, 305. Quantitative aspects of ascorbic acid metabolism in patients with primary hyperoxaluria.

BAKER, N. and SCHOTZ, M. C. (1964). *J. Lipid Res.*, **5**, 188. Use of multicompartmental models to measure rates of triglyceride metabolism in rats.

BERGNER, P. E. E. (1965). *Science, N.Y.*, **150**, 1048. Exchangeable mass: determination without assumption of isotopic equilibrium.

BERMAN, M. and SCHOENFELD, R. (1956). *J. appl. Phys.*, **27**, 1361. Invariants in experimental data on linear kinetics and the formulation of models.

BLEEHAN, N. M. and FISHER, R. B. (1954). *J. Physiol.*, **123**, 260. The action of insulin in the isolated rat heart.

BORN, G. V. R. and BÜLBRING, E. (1956). *J. Physiol.*, **131**, 690. The movement of potassium between smooth muscle and the surrounding fluid.

BROWNELL, G. L., BERMAN, M. and ROBERTSON, J. S. (1968). *Int. J. appl. Radiat. Isotopes*, **19**, 249. Nomenclature for tracer kinetics.

CUMMINGS, A. J., KING, M. L. and MARTIN, B. K. (1967). *Br. J. Pharmac. Chemother.*, **29**, 150. A kinetic study of drug elimination: the excretion of paracetamol and its metabolites in man.

DAYTON, S. (1959). *Circulation Res.*, **7**, 468. Turnover of cholesterol in the artery walls of normal chickens.

GEST, H., KAMEN, M. D. and REINER, J. M. (1947). *Archs Biochem. Biophys.*, **12**, 273. The theory of isotope dilution.

HALLBERG, D. (1965). *Acta physiol. scand.*, **64**, 306. Studies on the elimination of exogenous lipids from the blood stream.

HART, H. E. (1955). *Bull. math. Biophys.*, **17**, 87. Analysis of tracer experiments in non-conservative steady state systems.

HART, H. E. (1957). *Bull. math. Biophys.*, **19**, 61. Analysis of tracer experiments: II. Non-conservative non-steady state systems.

HEVESEY, G. and HAHN, L. (1940). *Biol. Meddr.*, **15** (5), 1. Turnover of lecithin, cephalin and sphingomyelin.

LAX, L. C., SIDLOFSKY, S. and WRENSHALL, G. A. (1956). *J. Physiol.*, **132**, 1. Compartmental contents and simultaneous transfer rates of phosphorus in the rat.

LONDON, I. M. and RITTENBERG, D. (1950). *J. biol. Chem.*, **184**, 687. Deuterium studies in normal man. I. The rate of synthesis of serum cholesterol. II. Measurement of total body water and water absorption.

PATERSON, J. Y. F. and HARRISON, F. A. (1968). *J. Endocr.*, **40**, 37. The specific activity of plasma cortisol in sheep after intravenous infusion of 1,2-^{3}H-cortisol, and its relation to the distribution of cortisol.

PRENTICE, T. C., SIRI, W. and JOINER, E. E. (1952). *Am. J. Med.*, **13**, 668. Quantitative studies of ascitic fluid circulation with tritium-labelled water.

REINER, J. M. (1953). *Archs Biochem. Biophys.*, **46**, 53. The study of metabolic turnover rates by means of isotopic tracers, I. Fundamental relations.

RESCIGNO, A. (1954). *Biochim. biophys. Acta*, **15**, 340. A contribution to the theory of tracer methods.

RESCIGNO, A. (1956). *Biochim. biophys. Acta*, **21**, 111. A contribution to the theory of tracer methods. Part II.

RITTENBERG, D. and FOSTER, G. L. (1940). *J. biol. Chem.*, **133**, 737. A new procedure for quantitative analysis by isotope dilution, with application to the determination of amino acids and fatty acids.

ROBERTSON, J. S. (1957). *Physiol. Rev.*, **37**, 133. Theory and use of tracers in determining transfer rates in biological systems.

SCHOTZ, M. C., BAKER, N. and CHAREZ, M. N. (1964). *J. Lipid Res.*, **4**, 569. Effect of carbon tetrachloride ingestion on liver and plasma triglyceride turnover rates.

SHARNEY, L., WASSERMAN, L. R. and GEVIRTZ, N. R. (1964). *Am. J. med. Electron.*, **3**, 249. Representation of certain mammillary n-pool systems by two-pool models.

SHARNEY, L., WASSERMAN, L. R., GEVIRTZ, N. R., SCHWARTZ, L. and TENDLER, D. (1965). *Am. J. med. Electron.*, **4**, 95. Significance of the time lag in 'tracer' movement: representation of unidirectionally-connected pool sequences by time lag.

SHARNEY, L., WASSERMAN, L. R., SCHWARTZ, L. and TENDLER, D. (1963). *Ann. N.Y. Acad. Sci.*, **108**, 230. Multiple pool analysis as applied to erythro-kinetics.

SHEPPARD, C. W. (1948). *J. appl. Phys.*, **19**, 70. The theory of the study of transfers within a multi-compartment system using isotopic tracers.

SHEPPARD, C. W. (1962). 'Basic Principles of the Tracer Method'. (John Wiley and Sons: New York.)

SHEPPARD, C. W. and HOUSEHOLDER, A. S. (1951). *J. appl. Phys.*, **22**, 510. The mathematical basis of the interpretation of tracer experiments in closed steady-state systems.

SLAYMAN, C. W. and TATUM, E. L. (1965). *Biochim. biophys. Acta*, **102**, 149. Potassium transport in *Neurospora*. II. Measurement of steady-state potassium fluxes.

SOLOMON, A. K. (1949). *J. clin. Invest.*, **28**, 1297. Equations for tracer experiments.

SOLOMON, A. K. (1953). *Adv. biol. med. Phys.*, **3**, 65. The kinetics of biological processes. Special problems connected with the use of tracers.

SOLOMON, A. K. (1960). Compartmental methods of kinetic analysis. *In* COMAR, C. L. and BRONNER, F. (Eds.) 'Mineral Metabolism'. Vol. IA, Chapter 5. (Academic Press: London.)

WRENSHALL, G. A. (1955). *Can. J. Biochem. Physiol.*, **33**, 909. Working basis for the tracer measurement of transfer rates of a metabolic factor in biological systems containing compartments whose contents do not intermix rapidly.

ZILVERSMIT, D. B., ENTENMAN, C. and FISHLER, M. C. (1943). *J. gen. Physiol.*, **26**, 325. On the calculation of turnover time and turnover rate from experiments involving the use of labelling agents.

CHAPTER FOUR

Systems Not in a Steady State

4.1 Introduction

Most systems which have been investigated using compartmental analysis have been in a steady state, and, apart from drug kinetics, little attention has been paid to other systems. This chapter will be concerned with a few types of systems which are not in a steady state because the size of the compartment is changing. The simple mathematical model of Figure 3.2 (equation 3.4) is restated,

$$\frac{d(Q_1)}{dt} = R_{10} - k_{01} Q_1 \qquad (4.1)$$

If R_{01} and k_{01} are constant but $R_{01} \neq k_{01} Q_1$ then the value of Q_1 i.e. the compartment size, will vary with time. If tracers are not present then equation 4.1 is a linear differential equation with Q_1 as the dependent variable and the relevant sections of Chapter 3 will apply. If tracer is present then equation 3.5 may be written in the form

$$\frac{d(q_1)}{dt} = -k_{01} Q_1 a_1$$

in which case both Q_1 and a_1 vary with time and the theoretical treatment becomes more complex.

4.2 Kinetics of foreign substances

This aspect of compartmental analysis became of great importance to pharmacologists long before radioactive tracers were employed for the study of metabolism. As examples we have the papers of Widmark (1920), Gehlen (1933), and the two classic papers of Teorell (1937a, b).

Widmark injected acetone into rabbits either subcutaneously or intraperitoneally. He measured the subsequent acetone concentrations in the blood and showed that the falling part of the curve fitted a single exponential function

$$q = q(0)\,e^{-\lambda t}$$

By fitting his data to this function he calculated that the volume of distribution of acetone in the rabbit $(q(0)/q)$ was 82–99% of body weight, and that the rate constant for the elimination of the compound (λ) was $0{\cdot}0008$–$0{\cdot}00215$ min.$^{-1}$. His first figure is too high because his system really consists of two compartments and he ignored the first compartment, the site of injection. He also obtained, and used, theoretical expressions for multiple dosage and for accumulation of material in a compartment.

Gehlen derived some theoretical expressions for what would now be considered a two compartment system.

Teorell, in his two papers, gave general treatments for the transfer of drugs between organs or tissues within an animal. For example, he considered the transfer from a site of administration into the blood, elimination from the body by excretion in the urine, exchange between tissues and destruction by a tissue. He used linear differential equations to describe these processes and solves the equations for a general model.

During the formulation of theoretical models for the study of the distribution and elimination of drugs (or other non-metabolised materials), it is implied that the body water, throughout which the drug is uniformly distributed, is the substance in a steady state within the compartments.

In most systems that have been studied the drugs appear to follow linear kinetics of the form that is applicable to tracers i.e. the rate of flow from a compartment is directly proportional to the concentration of drug within the compartment. Thus any of the earlier theoretical work, mentioned above, can be applied directly to the kinetics of tracers, and similarly the models and methods of Chapter 3 can be used in studying drug kinetics.

Wagner (1967) provides a recent example of the use of the

F

techniques of compartmental analysis in a study of the kinetics of a tetracycline drug. The compound was given orally to a subject and its subsequent concentration in serum was measured over a period of 16 hr. He chose to fit the data to the two compartment

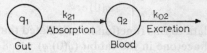

Fig. 4.1 Theoretical model to explain the absorption of a tetracycline drug from the gut into the blood, and excretion from the blood into the urine.

model of Figure 4.1 in which there is a short time lag between administration and absorption of the drug. Using the techniques of Chapter 6, the data was first plotted graphically to yield a function

$$q_2 = 2 \cdot 4 \, e^{-0.130t} - 3 \cdot 0 \, e^{-0.810t}$$

then the data were fitted by a digital computer program, incorporating the time lag, to give the more precise function

$$q_2 = 2 \cdot 65 \, (e^{-0.149 \, (t - 0.421)} - e^{-0.716 \, (t - 0.421)})$$

The fitted curve and the experimental data are shown in Figure 4.2.

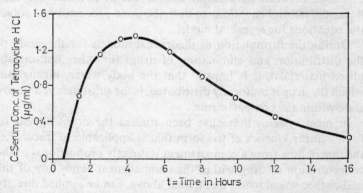

Fig. 4.2 Curve produced by a digital computer while fitting data for the concentration of a tetracyline drug in serum after an oral dose to the model of Figure 4.1 (after Wagner 1967). (Reproduced with the kind permission of the editor and publishers of *Clinical Pharmacology and Therapeutics*.)

Thus the time delay is calculated to be 0·42 hr., and the two rate constants for absorption and excretion are 0·72 hr.$^{-1}$ and 0·15 hr.$^{-1}$, respectively.

This direct utilisation of tracer kinetics is not restricted to drugs but can be applied to any foreign substance. After the intravenous injection of such compounds as mannitol, inulin, and thiosulphate, the rate of their elimination has been measured and shown to follow linear kinetics. Usually, the object of such experiments is to determine body fluid sizes and rates of clearance by the kidneys.

An example of this type of experiment is provided by Zender, Denkinger, and Falbriard (1965). They have used inulin as a tracer for the movement of water within the rabbit, and a model of the system is given in Figure 4.3. Compartment 1 represents the blood

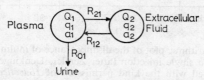

Fig. 4.3 A two compartment model to explain the exchange of water between plasma and extracellular fluid and the excretion of water as urine, using inulin as tracer.

plasma, compartment 2 is the extracellular fluid (ECF), $R_{21} = R_{12}$ is the exchange rate of water between the two compartments, and R_{01} is the rate of excretion by the kidneys. This model is similar to that of Figure 3.21 and the solution of the mathematical equations for compartment 1 (cf. equation 3.15) gives

$$q_1 = X_1 e^{-\lambda_1 t} + X_2 e^{-\lambda_2 t}$$

A single intravenous injection of inulin was given to a rabbit and subsequently the concentration of inulin within the plasma was determined. The experimental data were then fitted graphically to a sum of two exponentials (see Section 6.3). Figure 4.4 shows the results for one experiment. X_1, X_2, λ_1 and λ_2 were determined from

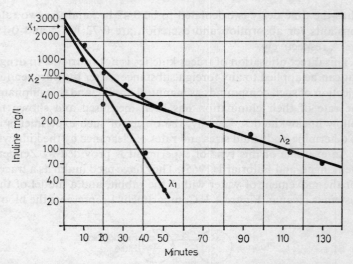

Fig. 4.4 Semi-logarithmic plot of the disappearance of inulin from the plasma of a rabbit after a single injection (after Zender, Denkinger, and Falbriard 1965). (Reproduced with the kind permission of *Helvetica Physiologica et Pharmacologica Acta.*)

this analysis, and hence the parameters of the original model could be calculated. The mean of seven experiments gave

$$q_1 \text{ (plasma)} = 8 \cdot 57 \pm 1 \cdot 73\% \text{ of body weight}$$
$$q_2 \text{ (ECF)} = 9 \cdot 43 \pm 2 \cdot 77\% \text{ of body weight}$$
$$R_{21} = R_{12} = 10 \cdot 85 \pm 4 \cdot 21 \text{ ml/min.}$$
$$R_{01} = 11 \cdot 33 \pm 2 \cdot 53 \text{ ml/min.}$$

This latter figure also provides an estimate for the inulin clearance rate for the rabbit. Note that for humans the accepted values for plasma and ECF volumes are about 5% and 16%, respectively.

Bray, Thorpe, and White (1951) describe a system in which a foreign compound after injection into an animal is excreted partly unchanged and partly as a less toxic material. Their system for the metabolism of toluene by the rabbit is shown in Figure 4.5.

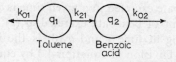

Fig. 4.5 A two compartment system in which toluene may be excreted unchanged (rate constant k_{01}), or else it may be oxidized to benzoate (rate constant k_{21}) and the latter excreted (rate constant k_{02}).

Toluene in compartment 1 was excreted unchanged (k_{01}) and was oxidized (k_{21}) to form benzoate present in compartment 2, and the latter was excreted, partly in the form of conjugates (k_{02}). In their experiments they measured the cumulative total amount of benzoate excreted in all forms i.e. they treated compartment 2 as an accumulation compartment. Their equations in our terminology are

$$\frac{d(q_1)}{dt} = -(k_{21}+k_{01})q_1 \qquad (4.2)$$

$$\frac{d(q_2)}{dt} = k_{21}q_1 \qquad (4.3)$$

and their results on one rabbit are shown in Figure 4.6. The total

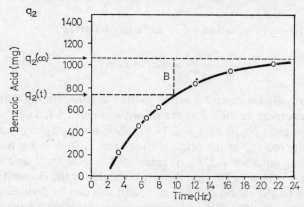

Fig. 4.6 Increase in the amount of benzoic acid excreted by a rabbit after the administration of toluene (from Bray, Thorpe, and White 1951).

amount of benzoate excreted in a given time is $q_2(t)$, $q_2(\infty)$ is the final total amount of benzoate excreted, $q_1(t)$ is the amount of toluene remaining at time t and $q_1(0)$ is the initial amount of toluene injected. Division of equation 4.3 by equation 4.2 gives

$$\frac{d(q_2)}{d(q_1)} = -\frac{k_{21}}{(k_{21}+k_{01})}$$

Integrating and applying boundary conditions i.e. when $t = t$, $q_1 = q_1$, $q_2 = q_2$, and when $t = \infty$, $q_1 = q_1(\infty)$, $q_2 = q_2(\infty)$. Then,

$$\frac{q_2(\infty)-q_2}{q_1} = \frac{k_{21}}{(k_{21}+k_{01})} \qquad (4.4)$$

Also, if $t = 0$

$$\frac{q_2(\infty)}{q_1(0)} = \frac{k_{21}}{(k_{21}+k_{01})} \qquad (4.5)$$

In equation 4.4, if $q_2(\infty)-q_2(t) = B$, then

$$k_{21} = \frac{(k_{21}+k_{01})B}{q_1} \qquad (4.6)$$

Substitution of equation 4.6 in equation 4.3 gives

$$\frac{d(q_2)}{dt} = (k_{21}+k_{01})B \qquad (4.7)$$

Their results for a second rabbit plotted in the form $\log B$ against time (compare Section 3.4) are shown in Figure 4.7. The slope of this line gives $(k_{21}+k_{01}) = 0\cdot14$ hr.$^{-1}$. Knowing $q_2(\infty) = 1\cdot09$ g benzoate (63·5% of the original dose) and $q_1(0) = 1\cdot3$ g toluene, and using equation 4.5, k_{21} is calculated as $0\cdot09$ hr.$^{-1}$ and k_{01} as $0\cdot05$ hr.$^{-1}$. This means that the rate constant for the elimination of toluene by the rabbit is $0\cdot14$ hr.$^{-1}$, and this can be divided into a rate constant of $0\cdot09$ hr.$^{-1}$ representing oxidation to benzoate, and one of $0\cdot05$ hr.$^{-1}$ representing excretion unchanged.

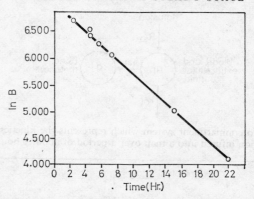

Fig. 4.7 Excretion of benzoic acid after toluene administration plotted semi-logarithmically according to equation 4.7 (after Bray, Thorpe, and White 1951). (Reproduced with the kind permission of the authors and the Editorial Board of the *Biochemical Journal*.)

4.3 Displacement of systems from a steady state

A few systems have been studied in which metabolites, normally present and in a steady state, have been displaced from an equilibrium concentration, and the subsequent variation in concentration has been determined, either while reaching a new steady state or while returning to the previous steady state. Provided that the displacement is not too large, the kinetics are linear. As an illustration of this statement, Hallberg (1965) injected a triglyceride emulsion into the blood stream of human subjects and measured its rate of removal. At low triglyceride concentrations the rate followed linear kinetics, but at higher concentrations the rate became independent of concentration i.e. followed zero order kinetics.

Tashjian and Whedon (1963) infused citrate solutions into human subjects and measured both the rise towards a new steady state during infusion, and the return to the previous steady state after cessation of infusion. Their theoretical model is shown in Figure 4.8. Compartment 1 comprises plasma plus extracellular fluid and compartment 2 represents bone and the viscera. The

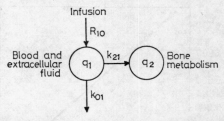

Fig. 4.8 A two compartment system which represents the kinetics of citrate solution infused into a man over a period of about 1 hour.

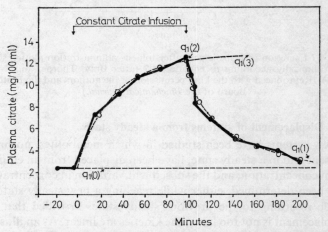

Fig. 4.9 Analysis of the infusion of citrate into a human subject. Constant infusion lasted from 0 to 90 min. ●——● observed values, O - - - O fitted values (after Tashjian and Whedon 1963). (Reproduced with the kind permission of the authors and the publishers of the *Journal of Clinical Endocrinology and Metabolism*.)

results of one of their experiments are shown in Figure 4.9. The values for citrate concentration in plasma obtained during their experiments fitted the following equations.

$$q_1(t) = q_1(3) - (q_1(3) - q_1(0)) e^{-(k_{21} + k_{01}) t} \qquad (4.8)$$

After infusion—

$$q_1(t) = q_1(1) - (q_1(2) - q_1(1)) e^{-(k_{21} + k_{01})} \qquad (4.9)$$

It is assumed that very little citrate is transfered from compartment 2 to compartment 1 so that a rate constant k_{12} can be neglected. However, it cannot be zero because $q_1(1)$ does not equal $q_1(0)$ i.e. the two curves of Figure 4.9 would probably consist of two exponentials if measurements were taken for long enough.

Data obtained from ten normal subjects were fitted to equations 4.8 and 4.9 using a digital computer program, and the mean of the results were

$$q_1(0) = 2\cdot4 \pm 0\cdot18 \text{ mg/100 ml}$$
$$q_1(1) = 3\cdot5 \pm 0\cdot30 \text{ mg/100 ml}$$
$$k_{21} + k_{01} = 0\cdot0365 \pm 0\cdot0022 \text{ min.}^{-1}$$

The values for $q_1(2)$ and $q_1(3)$ vary more widely depending on the subjects. Analysis of the urinary excretion of citrate gave a value for $k_{01} = 0\cdot0046$ min.$^{-1}$, therefore $k_{21} = 0\cdot0319$ min.$^{-1}$. The volume of compartment 1 is given by

$$\frac{100 \times R_{10}}{(\text{body weight})(k_{21}+k_{01})(q_1(3)-q_1(0))} \text{ ml/kg body weight}$$
$$= 193 \pm 13 \text{ ml/kg}$$

The citrate content of compartment 1 is

$$\frac{q_1(0) \times \text{volume}}{100} \text{ mg/kg} = 4\cdot7 \pm 0\cdot26 \text{ mg/kg}$$

The total turnover rate is

$$\text{compartment size} \times (k_{21}+k_{01}) \text{ mg/kg/day} = 238 \pm 8 \text{ mg/kg/day}$$

Note that many of the parameters for compartment 1 could not be determined without the separate analysis of urinary excretion, and even then, some of the parameters for compartment 2 remained unknown.

4.4 Systems with tracer present

In this type of system, not only is the amount of tracer q_i in a compartment i varying with time, but also the amount of substance Q_i is no longer constant.

If one has a general system containing n interconnected compartments, the total rate at which material flows into compartment i is

$$\sum R_{ij} \quad (j = 1, 2, \ldots n; i \neq j)$$

Similarly, the rate of flow out of the compartment is

$$\sum R_{ji} \quad (j = 1, 2, \ldots n; i \neq j)$$

The net rate at which material accumulates in compartment i is the difference between these two rates

$$\frac{d(Q_i)}{dt} = \sum R_{ij} - \sum R_{ji} \tag{4.10}$$

The kinetics for the flow of tracer into and out of the compartment are given by

$$\frac{d(a_i Q_i)}{dt} = \frac{d(q_i)}{dt} = \sum a_j R_{ij} - \sum a_i R_{ji}$$

$$= \sum a_j R_{ij} - a_i \sum R_{ji} \tag{4.11}$$

Multiplying equation 4.10 by a_i gives

$$a_i \frac{d(Q_i)}{dt} = a_i \sum R_{ij} - a_i \sum R_{ji} \tag{4.12}$$

and thence substituting equation 4.12 in equation 4.11

$$\frac{d(q_i)}{dt} = \sum a_j R_{ij} - a_i \sum R_{ij} + a_i \frac{d(Q_i)}{dt} \tag{4.13}$$

If we now differentiate the relationship

$$q_i = a_i Q_i$$

we obtain

$$\frac{d(q_i)}{dt} = a_i \frac{d(Q_i)}{dt} + Q_i \frac{d(a_i)}{dt} \tag{4.14}$$

From equations 4.13 and 4.14 we have finally,

$$Q_i \frac{d(a_i)}{dt} = \sum a_j R_{ij} - a_i \sum R_{ij}$$

or $$Q_i \frac{d(a_i)}{dt} = \sum R_{ij}(a_j - a_i) \qquad (4.15)$$

Goldwater and Stetten (1947) and Popják and Beeckmans (1950) have used equation 4.15 although not exactly in the form presented here. They have considered the system in which growing foetuses have synthesized cholesterol, and for this process some of the hydrogen atoms are derived from the water of the body fluids. The system is shown in Figure 4.10. At a particular time during the

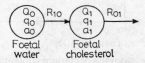

Foetal   Foetal
water   cholesterol

Fig. 4.10 A two compartment model in which deuterium oxide in compartment 1, kept at a constant enrichment, is used during the synthesis of foetal cholesterol in compartment 2.

development of the foetus the water within the foetus was labelled with deuterium and the deuterium concentration (a_0) was then kept at a constant value. The deuterium content of the foetal cholesterol was determined at various times and the results of Popják and Beeckmans when plotted are as in Figure 4.11. Adapting the general equation 4.15 to this particular problem produces

$$Q_1 \frac{d(a_1)}{dt} = R_{10}(a_0 - a_1) \qquad (4.16)$$

All the quantities Q_1, $d(a_1)/dt$, R_{10}, and a_1 varied with time. Q_1 and a_1 were determined directly for each particular time, $d(a_1)/dt$ was obtained by drawing tangents to the curve of Figure

78 MODELS FOR BIOLOGICAL SYSTEMS

4.11, and a_0, the maximum value reached by a_1, was estimated.
Equation 4.16 can be rearranged

$$R_{10} = \frac{Q_1 \, d(a_1)/dt}{(a_0 - a_1)}$$

and hence the rate of synthesis of cholesterol at a particular time
could be calculated.

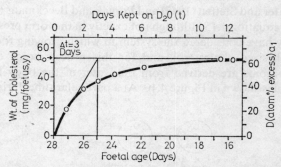

Fig. 4.11 The relation between the deuterium content of foetal cholesterol and
the number of days the mother was fed on D_2O (after Popják and Beeckmans
1950). (Reproduced with the kind permission of the authors and the Editorial
Board of the *Biochemical Journal*.)

Goldwater and Stetten calculated that for the rat half the
cholesterol present in the foetus at 20 days *in utero* had been
synthesized by the foetus within the previous 2·5 days. Popják
and Beeckmans showed that for the rabbit at 28 days *in utero* all
the foetal cholesterol was being synthesized within the foetus.
Also they calculated that the total cholesterol present was 55 mg
and the total rate of synthesis was 18·3 mg/day, of which 1·64
mg/day were synthesized by the liver.

REFERENCES

BRAY, H. G., THORPE, W. V. and WHITE, K. (1951). *Biochem. J.*, **48**, 88.
Kinetic studies of the metabolism of foreign organic compounds. I.
GEHLEN, W. (1933). *Arch. exp. Path. Pharmak.*, **171**, 541. Wirkungs-
stärke intravenös verabreichter Arzneimittel als Zeitfunktion.

GOLDWATER, W. H. and STETTEN, D. (1947). *J. biol. Chem.*, **169**, 723. Studies in fetal metabolism.

HALLBERG, D. (1965). *Acta physiol. scand.*, **64**, 306. Studies on the elimination of exogenous lipids from the blood stream.

POPJÁK, G. and BEECKMANS, M-L. (1950). *Biochem. J.*, **46**, 547. Synthesis of cholesterol and fatty acids in foetuses and in mammary glands of pregnant rabbits.

TASHJIAN, A. H. and WHEDON, G. D. (1963). *J. clin. Endocr. Metab.*, **23**, 1029. Kinetics of human citrate metabolism: studies in normal subjects and in patients with bone disease.

TEORELL, T. (1937a). *Archs int. Pharmacodyn. Thér.*, **57**, 205. Kinetics of distribution of substances administered to the body. I. The extravascular modes of administration.

TEORELL, T. (1937b). *Archs int. Pharmacodyn. Thér.*, **57**, 226. Kinetics of distribution of substances administered to the body. II. The intravascular modes of administration.

WAGNER, J. G. (1967). *Clin. Pharmac. Ther.*, **8**, 201. Use of computers in pharmacokinetics.

WIDMARK, E. M. P. (1920). *Acta med. scand.*, **52**, 87. Studies in the concentration of indifferent narcotics in blood and tissues.

ZENDER, R., DENKINGER, E. and FALBRIARD, A. (1965). *Helv. physiol. pharmac. Acta*, **23**, 199. Décroissance plasmatique de substances glomérulaires chez le lapin.

CHAPTER FIVE

Simulation Techniques

5.1 Introduction

Many biological, chemical, or physical systems are described by similar forms of mathematical equations, and the form of the equations for two particular systems taken from different fields may be identical. If so, the two systems are said to be analogues of each other and the behaviour of a variable in one system will be similar to that of the corresponding variable in the other system.

Most biological experiments using compartmental analysis are of several hours or days duration, but if a suitable physical analogue is available, then a comparable experiment may last only minutes or even milliseconds. Not only is there a considerable saving in the time taken to perform an experiment, but often the need to solve a set of simultaneous differential equations for each model tested, is avoided. It is obvious that, provided the analogue can be easily manipulated, it is often much more convenient and quicker to investigate the analogue situation before studying the biological system. In this chapter we will discuss the use of hydrodynamic models and analogue and digital computers for simulating the behaviour of tracers in compartmental systems.

5.2 Hydrodynamic models

Only very simple equipment, such as is available in any laboratory, is used in this technique. It can easily be set up and the results are readily observable. Apart from its occasional use in more serious work, it is an ideal technique for demonstrations to, or use by, students.

In these models compartments are represented by vessels (large beakers or flasks) containing water which represents the traced substance, so that the amount of water in the vessel corresponds to the amount of substance in the compartment. The vessels should be stirred to give rapid mixing i.e. so that the compartments are homogeneous. Transfer of water between the compartments is achieved either by gravity or by pumps. At zero time a concentrated dye solution is added to one of the vessels, then at suitable times thereafter samples can be taken from any of the compartments and the concentration of dye can be determined colorimetrically. The amount of dye added to the system represents the total amount of tracer added, and the concentration at a given place and time corresponds to specific activities (atom per cent, etc.). Figure 5.1 shows the theoretical model and a hydrodynamic

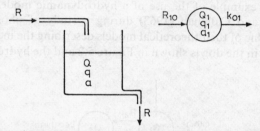

Fig. 5.1 A hydrodynamic model representing a one compartment system. The volume of water in the vessel (Q) represents the compartment size (Q_1), the amount of dye in solution (q) represents the amount of tracer in the compartment (q_1), and the dye concentration (a) represents the specific activity of the tracer (a_1). Input and output flow rates of water are equal (R) and represent the turnover rates of the material in the compartment (R_{10} and $Q_1 . k_{01}$).

model for a one compartment system. Q is the volume of water in the vessel (ml), q is the amount of dye present at a given time (g). Under steady state conditions, if R is the constant rate of inflow (ml/min.), the mathematical equation for the concentration of the dye (a in g/ml) is

$$\frac{d(q)}{dt} = -Ra$$

The rate of outflow depends on the head of water in the vessel, and therefore on Q for a parallel sided vessel

i.e. $R = kQ$

where k is a constant (min.$^{-1}$).

Therefore, $\dfrac{d(q)}{dt} = -kQa$

Comparison with the one compartment biological model (Section 3.4, equation 5.1) indicates the similarities of the two systems and the analogous behaviour of the corresponding variables.

$$\frac{d(q_1)}{dt} = -k_{01} Q_1 a_1 \qquad (5.1)$$

A good example of the use of a hydrodynamic model is given by Shore and Callahan (1963) during a study of phospholipid kinetics. One of their theoretical models describing the metabolism of lecithin in the dog is shown in Figure 5.2 and the hydrodynamic

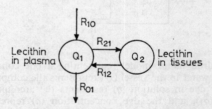

Fig. 5.2 A two compartment model to represent the kinetics of the distribution of lecithin between plasma and other tissues.

model is presented in Figure 5.3. The input R_{10} represents synthesis of new material within compartment 1 and the output R_{01} represents removal of material by degradation. R_{12} and R_{21} represent exchange of phospholipid between the tissue (1) and plasma (2) compartments. The dye used was bromsulfalein (BSP) and the change of BSP concentration during their experiments is shown in Figure 5.4. Table 5.1 compares the measured and the calculated

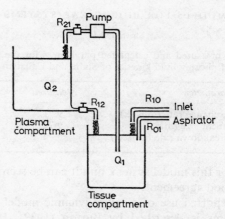

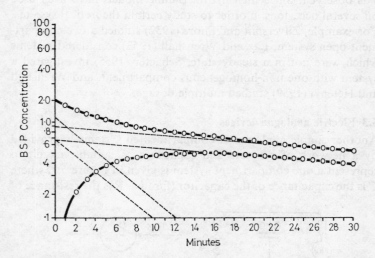

Fig. 5.3 The hydrodynamic system used by Shore and Callahan to simulate the model of Figure 5.2 (after Shore and Callahan 1963). (Reproduced with the kind permission of the authors and the New York Academy of Sciences.)

Fig. 5.4 Changes in the concentration of dye in the two compartments of the hydrodynamic model of Figure 5.3 after addition of dye to compartment 1 (after Shore and Callahan 1963). (Reproduced with the kind permission of the authors and the New York Academy of Sciences.)

TABLE 5.1

Comparison of measured and calculated parameters for the hydrodynamic
model of phospholipid kinetics of Shore and Callahan (1963).

Parameter	Measured	Calculated
Plasma–tissue exchange rate (ml/sec.)	9·71	10·1
Tissue compartment size (litres)	9·72	9·81
Synthesis and breakdown rate (ml/sec.)	5·72	5·70

parameters for this model system, and it can be seen that there is
reasonably good agreement.

The first reported use of a hydrodynamic model in compartmented systems is described by Burton (1939). He used this
technique to study a two compartment system originally in a steady
state. The system was displaced from equilibrium by a sudden
change in the flow rates and then the return to a new steady state
was observed. Since then hydrodynamic models have been used
on several occasions in order to study certain theoretical systems.
For example, Zilversmit and Shore (1952) studied a two compartment open system, Lax and Wrenshall (1953) considered systems
which were not in a steady state, Schacter (1955) investigated a
system with one non-homogeneous compartment, and Wrenshall
and Hetenyi (1959) studied multiple dosage.

5.3 Electric analogue devices

Another simple analogue of compartmented systems is provided
by a network of resistors and capacitors. A suitable circuit to
represent a one compartment system is given in Figure 5.5, where
C is the capacitance of the capacitor (farads), R is the resistance of

Fig. 5.5 An electric analogue to represent a one compartment system. The
amount of charge held by the capacitor (Q) represents the amount of tracer in
the system (q_1).

the resistor (ohms), and V (volts) is the voltage across the capacitor at some time t after the relay has been opened. If, also at time t, the charge in the system is Q and the current flowing through the resistor is I, then the voltage across the capacitor is Q/C and the voltage across the resistor is IR which equals $R[d(Q)/dt]$. Since the sum of voltages in the circuit must be zero

$$R\frac{d(Q)}{dt} + \frac{Q}{C} = 0$$

Hence

$$\frac{d(Q)}{dt} = -\frac{1}{RC}Q$$

Comparison with the equation for the tracer system

$$\frac{d(q_1)}{dt} = -k_{01}q_1$$

shows that the amount of charge is the analogue of the amount of tracer and that $1/RC$ corresponds to k_{01}.

Hickey and Brownell (1954) were probably the first to use an electric analogue device in compartmental analysis. They gave ^{131}I-iodide to men and used the model of Figure 5.6 to explain the

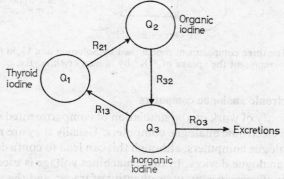

Fig. 5.6 The three compartment model used by Hickey and Brownell (1954) to represent the movement of ^{131}I in man.

subsequent kinetics of the tracer. The data were fitted to the exponential functions by the electric analogue computer and hence the parameters of the model could be calculated. For example, results on one subject gave: thyroidal iodine 7·85 mg, organic iodine 1·4 mg, inorganic iodine 52 μg, R_{21} 96 μg/day, and R_{03} 158 μg/day.

Solomon and Gold (1955) provide a second example. They allowed freshly drawn human erythrocytes to equilibrate with $^{42}K^+$, and after transfering the cells to unlabelled plasma, they measured the rate of release of tracer into the plasma. The data was fitted to a sum of three exponentials using an electric analogue device. Figure 5.7 illustrates the three compartment model which is most compatible with the data, out of three tested, and it also indicates the calculated parameters.

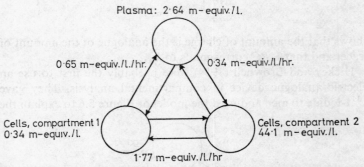

Fig. 5.7 The three compartment model used by Solomon and Gold (1955) to represent the uptake of $^{42}K^+$ by human erythrocytes.

5.4 Electronic analogue computers

The majority of work on the simulation of compartmented systems has used electronic analogue computers. Usually they are referred to as analogue computers, although this can lead to confusion with simpler analogue devices. In these machines voltage is used as the analogue of specific activity or quantity of tracer, and the voltages are operated upon by units termed operational amplifiers which are connected with other components in order to carry out the steps of addition, multiplication by a constant, or integration. In the following pages we will give a brief description of the components involved and how to program simple problems.

An *operational amplifier* is a high gain D.C. voltage amplifier. An amplifier is usually represented by the triangular symbol of Figure 5.8, and the relation between the input and output voltages is

$$V_0 = -MV_a$$

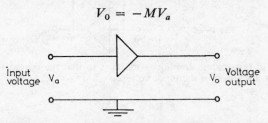

Fig. 5.8 A block diagram illustrating an operational amplifier with its input and output connections.

where M is called the 'gain' of the amplifier. Note that the sign of the voltage is reversed. Voltages are always measured with respect to earth, therefore it is usual to omit the earth line as in Figure 5.9.

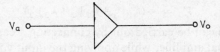

Fig. 5.9 The usual representation of an operational amplifier in which the earth connections are omitted from the diagram.

If our operational amplifier is connected so that the input voltage is applied through an input resistor of value R_i, and the output voltage is fed back to the input by a resistor of value R_f then we have the situation of Figure 5.10. It can be shown that the ratio of input to output voltages is now given by

$$\frac{V_0}{V_i} = -\frac{R_f M}{R_f + R_i (1+M)}$$

The value of M in operational amplifiers is usually of the order of 10^6 i.e. $R_f + R_i \ll R_i M$, so that

$$\frac{V_0}{V_i} \simeq - \frac{R_f}{R_i}$$

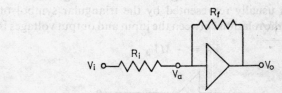

Fig. 5.10 An operational amplifier connected so that the output voltage is a multiple or a fraction of the input voltage.

$$V_0 = - \frac{R_f}{R_i} . V_i \quad \text{or} \quad V_0 = -k . V_i$$

If $R_f = R_i$, then $V_0 = -V_i$ and we have carried out a *sign reversal*. If $R_f / R_i = k$, then $V_0 = -kV_i$ and as well as reversing the sign we have *multiplied by a constant*.

Summation is achieved by applying the voltages to be added to the operational amplifier via separate input resistors as in Figure 5.11. If $R_f / R_i = k_i$, then the output voltage is given by

$$V_0 = -(k_1 V_1 + k_2 V_2 + k_3 V_3)$$

Integration can be carried out by connecting a capacitor across the operational amplifier while applying the input voltage through

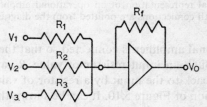

Fig. 5.11 An operational amplifier connected in order to perform additions. $V_0 = -(k_1 V_1 + k_2 V_2 + k_3 V_3)$, where

$$k_1 = \frac{R_f}{R_1}, \quad k_2 = \frac{R_f}{R_2}, \quad k_3 = \frac{R_f}{R_3}.$$

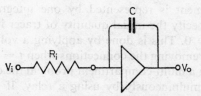

Fig. 5.12 An operational amplifier connected so that the input voltage is integrated to give the output voltage.

$$V_0 = -\frac{1}{R_i C}\int_0^t V_i \, . \, dt$$

a resistor, as in Figure 5.12. The output voltage is given by

$$V_0 = -\frac{1}{R_i C}\int_0^t V_i \, dt$$

or,

$$V_0 = -\frac{1}{T}\int_0^t V_i \, dt$$

where $T(= R_i C)$ is known as the time constant for the integration.

An alternative means for multiplying by a constant less than 1

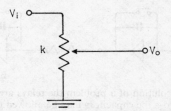

Fig. 5.13 A potentiometer connected so that the input voltage is multiplied by a constant whose value is less than 1. $V_0 = kV_i$.

is to use a potentiometer, as in Figure 5.13. Thus the output voltage V_0 is

$$V_0 = kV_i$$

The advantage of this method is that the value of k may be rapidly changed whenever desired.

When the analogue computer is used in compartmental analysis each compartment is represented by one integrator, and it is necessary to specify the initial quantity of tracer in the compartment when $t = 0$. This is done by applying a voltage across the capacitor and removing the connections when $t = 0$, thus starting the solution. In a multi-compartment problem the connections are all removed simultaneously by using a relay. If the amount of tracer in the compartment is to be zero then the capacitor is shorted by the relay, if tracer is inserted then some predetermined voltage is applied. The situations before and during solution are shown in Figures 5.14 and 5.15.

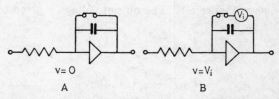

Fig. 5.14 Setting the initial conditions for an integrator. In A the voltage across the capacitor is zero, and in B the voltage has some predetermined value V_i.

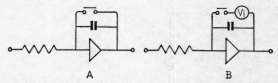

Fig. 5.15 During the solution of a problem the relays are opened so that the voltages across the capacitors may be allowed to change.

Solution of a differential equation. A one compartment system is represented by the mathematical model

$$\frac{d(a_1)}{dt} = -k_{01} a_1$$

For simulation, a_1 (say in counts/min.) is represented by a voltage V_1 (volts), and $a_1(0)$ by a voltage V_0. The ratio a_1/V_1 (counts/min./volt) is known as the scaling factor. The analogue computer is programmed as in Figure 5.16. The input to the integrator is $k_{01} V_1$. Integration would produce

$$\int_0^t \left(-\frac{d(V_1)}{dt} \right) dt$$

i.e. $-V_1$, but remembering that a sign change occurs on integration, the output is in fact $+V_1$. Applying this output to a potentiometer of value k_{01} yields a voltage $k_{01} V_1$ which is then used for the input. Starting conditions are introduced by applying a voltage of V_0 across the capacitor and opening the relay at $t = 0$.

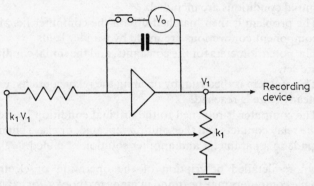

Fig. 5.16 An operational amplifier and its connections which would solve the differential equation $da_1/dt = -k_1 a_1$. i.e. one which would simulate a one compartment system.

Visualising the output is usually carried out in one of two ways. For one method the voltage V_1 is plotted directly using a chart recorder. In this method about 30–60 seconds are required for a solution. The alternative method is to display the output on an oscilloscope screen in which case the solution is obtained during the time taken for one sweep of the oscilloscope beam. After this time, usually a small fraction of a second, the solution is repeated in order to reinforce the oscilloscope trace. The difference in solution time is obtained by adjusting the integrator time constant; the larger the time constant the longer the time taken for solution.

The latter method, because of its rapidity, is most popular. A large number of values of k_{01} can be investigated in a very short time.

The order of operations necessary to simulate a problem are summarised as follows.

(1) The mathematical model is stated as a set of linear differential equations.
(2) A flow chart is drawn showing the necessary computer units and their relevant connections.
(3) Suitable scaling factors and integrator time constants are chosen.
(4) Initial conditions are included.
(5) The problem is then 'patched up' on the computer i.e. all the component connections are made by wander leads.
(6) The potentiometers for the constants, and the initial conditions are set.
(7) The problem is then run, by opening the relay contacts, until a steady state is reached.
(8) The computer is returned to the original condition by closing the relay contacts and the solution repeated, or else a change is made to a parameter and another solution obtained.

A more detailed description of the operation of electronic analogue computers may be found in many textbooks, for example, those of Stice and Swanson (1965) and Stewart and Atkinson (1967).

An example of the application of electronic analogue computers to compartmental analysis is that of Blomstedt and Plantin (1965). They injected ^{131}I-thyroxine intravenously into human subjects and measured its distribution in several compartments. Their first model is shown in Figure 5.17 where compartment 1 represents the intravascular component, compartment 2 the accumulated excretion, and compartment 3 the extravascular component. Their results, together with the analogue simulated curves are shown in Figure 5.18 and it is seen that the agreement is not very good. The model was then modified to that of Figure 5.19 where the additional compartment 4 represents an irreversible extravascular compartment. The results and the simulated curves are reproduced in

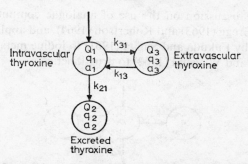

Fig. 5.17 The first model used by Blomstedt and Plantin (1965) in an attempt to explain the kinetics of ^{131}I-thyroxine in man. Compartment 1 represents the intravascular component, compartment 2 the accumulated excretion, and compartment 3 the extravascular component.

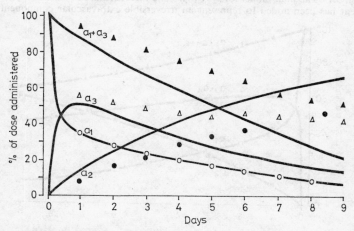

Fig. 5.18 The distribution of ^{131}I-thyroxine between the various compartments according to the model of Figure 5.17 (from Blomstedt and Plantin 1965). (Reproduced with the kind permission of the authors and the editor of *Acta Endocrinologica*.) — predicted by the analogue computer, ▲ $a_1 + a_3$, △ a_3, ○ a_1, ● a_2.

Figure 5.20 and it is obvious that they coincide well. The second model is, therefore, a more plausible description of the physiological system.

Further discussion on the use of analogue computers will be found in Gregg (1963) and Robertson (1964), and applications are described by Fukuda and Sugita (1961) to iodine metabolism and by Taylor and Wiegand (1962) to drug kinetics.

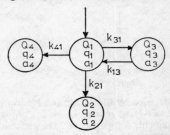

Fig. 5.19 The final model for ^{131}I-thyroxine metabolism. A fourth compartment has been added to represent an irreversible extravascular component.

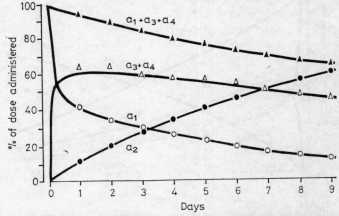

Fig. 5.20 The distribution of ^{131}I-thyroxine between the various compartments according to the model of Figure 5.19 (from Blomstedt and Plantin 1965). (Reproduced with the kind permission of the authors and the editor of *Acta Endocrinologica*.) Legend as for Figure 5.18 except ▲ $a_1 + a_3 + a_4$, △ $a_3 + a_4$.

5.5 Digital computers

Digital computers have as yet found little application for the simulation of compartmented systems. The reason is that the vast majority of work to date is on small linear systems and these are best simulated on analogue computers. Linear systems were defined in Section 3.1 as systems in which the rate of movement of a substance is equal to the product of the concentration of the substance and an appropriate rate constant. This condition applies to tracers in a steady state system, and to most systems containing foreign compounds. However, there are other systems in which this simple relationship does not obtain e.g. in non-steady state systems, where hormones may be having a regulatory effect, or where first order kinetics may not apply because a transport mechanism is partly or wholly saturated. These are non-linear systems, where an analytical solution is usually impossible, and digital computers are becoming increasingly useful in simulating these systems.

The principle behind the use of digital computers is that the differential equations are solved by numerical analysis i.e. given a set of initial conditions for the dependent variables, numerical values for each variable are calculated at successive times after time zero. By using suitable graph routines these values are then plotted in a graphical form using either a line printer or a graph plotter. Further solutions are obtained by changing the parameters and repeating the procedure. The simulation of multi-enzyme systems and ecological systems involves a similar procedure, although the differential equations in these problems are non-linear e.g. Chance *et al.* (1960) and Garfinkel, MacArthur, and Sack (1964).

The only instance known to the author of the use of the digital computer for the simulation of a linear problem in compartmental analysis is the investigation by Vuille (1965) of several models to describe the kinetics of distribution of iron in man. He has simulated several possible situations for each model by using varying values for the model parameters, and in some cases has compared the resulting curves with experimental data.

5.6 Comparison of analogue and digital computers

The advantage of analogue computers over digital computers lies principally in their faster operation. Analogue computers solve differential equations simultaneously whereas in digital computers each operation is carried out sequentially. In addition, digital computers require time for the print out of results. Analogue computers are also much easier to program and many solutions can be obtained in a very short time. They are also much cheaper to buy. The advantages of digital computers lies partly in their greater flexibility and capacity, and partly in their greater precision. The accuracy of analogue computers is of the order of 0.1% to 1% and as the number of operational amplifiers increases the overall accuracy decreases. On the other hand numerical analysis can be carried out with an accuracy greater than 1 in 10^6. The other advantage of digital computers is that they will handle non-linear situations as easily as linear ones, and they will also cope with complex systems that have very many variables to simulate.

REFERENCES

BLOMSTEDT, B. and PLANTIN, L. O. (1965). *Acta endocr., Copnh.*, **48**, 536. The extrathyroidal distribution of ^{131}I-thyroxine.

BURTON, A. C. (1939). *J. cell. comp. Physiol.*, **14**, 327. The properties of the steady state compared to those of equilibrium as shown in characteristic biological behaviour.

CHANCE, B., GARFINKEL, D., HIGGINS, J. and HESS, B. (1960). *J. biol. Chem.*, **235**, 2426. Metabolic control mechanisms. V. A solution for the equations representing interaction between glycolysis and respiration in ascites tumour cells.

FUKUDA, N. and SUGITA, M. (1961). *J. theor. Biol.*, **1**, 440. Mathematical analysis of metabolism using an analogue computer: I. Isotope kinetics of iodine metabolism in the thyroid gland.

GARFINKEL, D., MACARTHUR, R. H. and SACK, R. (1964). *Ann. N.Y. Acad. Sci.*, **115**, 943. Computer simulation and analysis of simple ecological systems.

GREGG, E. C. (1963). *Ann. N.Y. Acad. Sci.*, **108**, 128. An analog computer for the generalized multi-compartment model of transport in biological systems.

HICKEY, F. C. and BROWNELL, G. L. (1954). *J. clin. Endocr. Metab.*, **14**, 1423. Dynamic analysis of iodine metabolism in four normal subjects.

LAX, L. C. and WRENSHALL, G. A. (1953). *Nucleonics*, **11**(4), 18. Measurement of turnover rates in systems of hydrodynamic pools out of dynamic equilibrium.

ROBERTSON, J. S. (1964). *Ann. N.Y. Acad. Sci.*, **115**, 553. Analog computation: definition and characteristics.

SCHACTER, H. (1955). *Can. J. Biochem. Physiol.*, **33**, 940. Direct versus tracer measurement of transfer rates in a hydrodynamic system containing a compartment whose contents do not intermix rapidly.

SHORE, M. L. and CALLAHAN, R. (1963). *Ann. N.Y. Acad. Sci.*, **108**, 147. Application of hydrodynamic analogs and digital computer to the study of phospholipid kinetics.

SOLOMON, A. K. and GOLD, G. L. (1955). *J. gen. Physiol.*, **38**, 371. Potassium transport in human erythrocytes: evidence for a three compartment system.

STEWART, C. A. and ATKINSON, R. (1967). 'Basic Analogue Computer Techniques'. (McGraw-Hill: London.)

STICE, J. E. and SWANSON, B. S. (1965). 'Electronic Analog Computer Primer'. (Blaisdell Publishing Co.: New York.)

TAYLOR, J. D. and WIEGAND, R. G. (1962). *Clin. Pharmac. Ther.*, **3**, 464. The analog computer and plasma drug kinetics.

VUILLE, J-C. (1965). *Acta physiol. scand.*, **65**, S.253, 1. Computer simulation of ferrokinetic models.

WRENSHALL, G. A. and HETENYI, G. (1959). *Metabolism*, **8**, 531. Successive measured injections of tracer as a method for determining characteristics of accumulation and turnover in higher animals with access limited to blood: tests in hydrodynamic systems and initial observations on insulin action in dogs.

ZILVERSMIT, D. B. and SHORE, M. L. (1952). *Nucleonics*, **10** (10), 32. A hydrodynamic model of isotope distribution in living organisms.

CHAPTER SIX

Treatment of Experimental Data-Curve Fitting and Model Fitting

---

6.1 Introduction—some difficulties of curve fitting

For the types of problem considered in this book the solutions of the mathematical models will all be of the form

$$a_i = \sum X_j e^{-\lambda_j t} \quad (j = 1, 2, \ldots n) \tag{6.1}$$

where n is the number of terms or compartments. Perhaps $\lambda_1 = 0$, in which case the first term will be a constant term, X_1.

The experimental data will consist of a set of points (a_{ik}, t_k) and the problem is to fit these data to the expression 6.1 and to determine the best values of the parameters X_j and λ_j. The errors in the observations (usually restricted to a_{ik}) consist of two parts; errors due to analysis and errors due to biological variation during the course of the experiment. These errors are not usually separated and provided they are sufficiently small they approximate to a normal distribution. The best methods, therefore, for calculating the parameters involve the minimisation of a sum of squares function,

$$\text{sum of squares} = \sum (a_{ik} - \sum X_j e^{-\lambda_j t_k})^2 \tag{6.2}$$

$$(k = 1, 2, \ldots m)$$

where m is the number of observations of the variable a_i.

Mathematical expressions consisting of sums of exponentials occur in many branches of science, and the problem of fitting experimental data using the least squares function of equation 6.2 has been looked at by a wide variety of scientists over a long period of time. Many mathematical solutions have been published, but when they are applied to experimental data the difficulties can become enormous.

Some of the difficulties which arise in attempting to separate exponentials can be illustrated by reference to Myhill (1967) and to Glass and de Garreta (1967). Both of these papers are concerned with how the accuracy of estimating the parameters X_j, λ_j is affected by several factors. The more important of these are (1) the ratio λ_1/λ_2, (2) the error in the data, and (3) the number of data points taken. They have both investigated a two exponential function, and obviously their conclusions apply to this one particular case, but presumably they can be generalised to cover other similar functions.

Two terms $X_i\,e^{-\lambda_i t}$ and $X_j\,e^{-\lambda_j t}$ can only be separated if λ_i and λ_j are appreciably different. Normally they should differ by at least a factor of two. If they are too close in value they will be represented by only one term ('lumping', see Section 3.16). In Myhill's paper, for $\lambda_1/\lambda_2 = 2$, his data can only be fitted if the error in the data is 1% or less, and for a ratio $= 4$, the error must be 5% or less. Van Liew (1962) also illustrates this point by considering sums of exponentials in which their λ are very close together. Resolution into exponential terms can be achieved, but the fitted functions bear no similarity to the original functions used to generate the data.

The values of the estimated parameters, especially the λ, are usually very sensitive to small changes in the experimental data. This is best measured by comparing the estimated standard deviations of the calculated parameters with the error in the original experimental data. Thus Myhill shows that for a data error of 1% the parameter error can be as high as 14%, at 2% it can be 17%, at 5% up to 44%, and at 10% as high as 86%.

The other important factor which has been investigated is the effect of increasing the number of experimental data points. Glass and de Garreta, for example, show that by increasing the number of points from 11 to 21 the mean of the standard deviations of the estimated parameters is reduced by about 25–33%. Myhill, simi-

H

larly, shows that an increase from 11 to 31 data points reduces the standard deviations of the estimates by about 30%.

Another factor which has not been investigated thoroughly, but seems obvious, is that it is difficult to determine a value of λ if its reciprocal is either very small or very large compared with the time span of the experiment. To overcome this, it is therefore desirable to design the experiment so that the time scale, or the range of data points, is such that the data are collected at suitable times to allow the λ to be calculated.

The preceeding chapters would suggest that compartmental analysis allows one to investigate and analyze quite complex biological systems. However, in practice, the potentiality of the technique in fulfilling its theoretical promise depends ultimately on the ability of fitting experimental data to sums of exponentials.

It has already been demonstrated that the successful use of the technique depends initially on the design of good experiments which allow a large amount of highly accurate data to be collected. The number of points ought to be in excess of ten, preferably two or three times this quantity, and each data point ought to have a standard deviation of 1%, or better, if more than three exponentials are to be fitted. A reading of the original literature will show that the design of many investigations and the collection of data fall considerably short of the above recommendations, with the result that the calculated results are often very approximate.

Assuming that a large quantity of good quality data is available, the remainder of the chapter will be mainly devoted to the methods which are available for separating the exponential terms of equation 6.1.

For some investigations, it may be that the model for the system is unknown and all that is required is a mathematical expression relating the concentration of tracer in different parts of the system to time. The necessary number of exponential terms is not usually known in advance. In this case a fit is made to a small number of terms and then the procedure is repeated with increasing numbers of terms until no further improvement in fit is obtained. If the data are not very accurate there may be a choice between the number of terms possible, in which case it is usual to select the expression with the smallest number.

6.2 A note on the use of statistics

In any procedure for which values are calculated from experimental data it is desirable to have some estimate of the reliability of the results. Most investigators therefore carry out a statistical analysis using their data and usually calculate a standard deviation for their estimated parameters. In the case of compartmental analysis, the most common method for calculating the standard deviations is to use the information available within a given set of experimental data. Thus, in graphical analysis (Section 6.3) they can be calculated during the fitting of regression lines, and in the case of data fitting by a digital computer program (Section 6.5) they can be obtained from a variance–covariance matrix. If one refers to the sum of squares function (equation 6.2), it is obvious that the relations between the sum of squares of residuals and many of the parameters to be estimated are not linear. The most commonly used procedures of statistics apply only to linear relationships, hence if standard deviations are calculated as indicated above they will not be particularly reliable values. The more non-linear equation 6.2 becomes, the more approximate will be the estimated standard deviations. The only reliable method for estimating standard deviations is to repeat a given experiment several times and then calculate a set of parameters for each experiment. Several estimates are thus obtained for the individual parameters and hence a reliable mean and standard deviation can be calculated for each parameter.

6.3 Graphical analysis

Historically, this method was first used in radioactive techniques (1) to separate and determine the half-lives of a mixture of radio-nuclides, and (2) to study a radioactive series in which transformations occurred through a series of radionuclides until a stable end product was obtained. The first application to compartmental analysis appears to be by Cohn and Brues (1945).

The technique depends on the property of equation 6.1 in that as t becomes large the contributions of the terms $\sum X_j e^{-\lambda_j t}$ ($j = 1, 2, \ldots n-1$) in relation to the final term $X_n e^{-\lambda_n t}$ become negligible. Thus

$$a_i \approx X_n e^{-\lambda_n t}$$

Taking logarithms (compare the treatment of equation 3.8) we have

$$\log a_i = \log X_n - 0 \cdot 4343 \lambda_n t \qquad (6.3)$$

If the experimental data are plotted in logarithmic form against t, or more conveniently it may be plotted directly on semi-logarithmic graph paper, the straight line given by equation 6.3 can be fitted to the last few points. Thus, X_n can be calculated from the intercept when $t = 0$, and λ_n from the slope. The next step in the analysis is to subtract the term $X_n e^{-\lambda_n t}$ (or $\log X_n - 0 \cdot 4343 \lambda_n t$) from each of the data points not falling on the above straight line and the resulting curve will be represented by

$$a_i' = \sum X_j e^{-\lambda_j t} \qquad (j = 1, 2, \ldots n-1)$$

The fitting of straight lines is then repeated until all the terms have been separated.

This procedure is well illustrated by Schloerb $et\ al.$ (1950) who measured the kinetics of water movement in the human body using deuterium oxide as tracer. Their values of arterial deuterium oxide concentrations for 70 min. after an intravenous injection are given in Figure 6.1 together with the straight lines which have been separated by the above procedure. The fitted function is

$$q = 0 \cdot 390 e^{-0 \cdot 565\ t} + 0 \cdot 234 e^{-0 \cdot 0727\ t} + 0 \cdot 146$$

It consists of three exponential terms of which one is e^{-0}, this suggesting that there are three compartments present in the system. No attempt was made in this paper to analyze the system further, but it was suggested that the compartmentation might be caused by (1) transfer across cell membranes, and (2) transfer across capillaries.

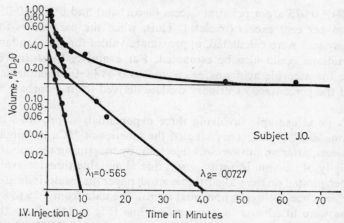

Fig. 6.1 Semi-logarithmic plot of serum deuterium concentration after a single injection of deuterium oxide in a man. The method of graphical analysis (from Schloerb *et al.* 1950). (Reproduced with the kind permission of the *Journal of Clinical Investigation*.)

This method of graphical analysis means that the most accurately determined parameters are those of the last term X_n, λ_n, and the accuracy of determination decreases with each successively estimated pair of constants. Partly for this reason and partly because of the time scale it is almost impossible to fit more than three or four terms by this technique.

In the foregoing the straight lines are fitted by eye. This means that neither are they the best possible fit, nor are any statistical calculations possible. An improvement is to fit linear regression lines to each straight line section during the analysis, so that the best estimate for the parameters, together with a standard deviation, are obtained.

Atkins *et al.* (1964) have used single regression lines for one-exponential functions. A subject was given a single oral dose of $1\text{-}^{13}C$-ascorbic acid and the isotopic enrichments of urinary ascorbic acid and urinary oxalic acid were measured at various times afterwards. The regression lines for one experiment had a mean slope of 0.028 ± 0.0028 day^{-1}, and the intercepts at $t = 0$ were

0.884 ± 0.075 atom per cent excess (ascorbate) and 0.309 ± 0.054 atom per cent excess (oxalate). Thus, when the parameters for the system were calculated, approximate values for the standard deviations could also be estimated. For example, the turnover rate for ascorbic acid was calculated as 0.44 ± 0.06 m-mole/day and the percentage of urinary oxalate derived from ascorbate as $35 \pm 7\%$.

A good example involving three exponentials is provided by Krane et al. (1956). They studied the kinetics of ^{45}Ca in human subjects, after an intravenous injection, by measuring the specific activity of calcium in serum and urine at suitable time intervals. When plotted on semi-logarithmic graph paper the results indicated the presence of three exponential terms. The data from 1·25 to 8·5 days were fitted first to a regression line (Fig. 6.2A) giving the results $X_3 = 13.38 \pm 0.97\%$ dose/g Ca and $\lambda_3 = 0.279 \pm 0.015$ day^{-1}. This line was subtracted from the data and a second regression line fitted to the residual values between $t = 0.13$ and 0.67 days (Fig. 6.2B). The calculated parameters were $X_2 = 28.21 \pm 1.88\%$ dose/g Ca and $\lambda_2 = 3.20 \pm 0.05$ day^{-1}. Similarly with the remaining points (Fig. 6.2C) they obtained $X_1 = 66.65 \pm 20.50\%$ dose/g Ca and $\lambda_1 = 36.60 \pm 9.31$ day^{-1}. Notice that the earlier parameters are estimated more accurately than the subsequent ones. No theoretical or mathematical model was proposed for their system so their experimental data were not analyzed further.

An extension of this procedure is valuable if observations have been made on more than one compartment. The regression values of X_n and λ_n can be obtained, as in the previous paragraphs, for each compartment and then the weighted means $\bar{X}_n$ and $\bar{\lambda}_n$ computed. Using these values $\bar{X}_n e^{-\bar{\lambda}_n t}$ can be subtracted from the data and the next pair of parameters $\bar{X}_{n-1}$ and $\bar{\lambda}_{n-1}$ calculated in a similar manner.

Apart from the simple case of fitting a regression line to a single exponential, the use of graphical analysis has become obsolescent. Its main use nowadays ought to be restricted to the estimation of initial values of parameters for subsequent use in one of the curve

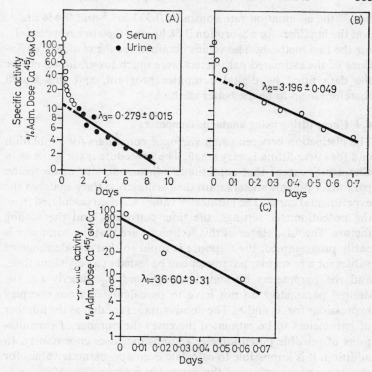

Fig. 6.2 Graphical analysis of the data for ^{45}Ca metabolism into three exponential components, illustrating the use of regression lines to calculate the values of X_i and λ_i (from Krane *et al.* 1956). (Reproduced with the kind permission of the *Journal of Clinical Investigation*.)

fitting procedures of Section 6.5 To illustrate this point Wagner and Metzler (1967) gave 500 mg erythromycin orally to a human subject and subsequently measured the concentration of the drug in the blood. The results were analyzed both graphically and by using a digital computer program, and the rate constants for absorption and elimination were calculated. The parameters estimated by the two procedures were substantially different. Thus, the absorption rate constants were 0·899 hr.$^{-1}$ and 0·708 hr.$^{-1}$, the elimination rate constants 0·393 hr^{-1} and 0·444 hr.$^{-1}$, and the lag times for absorption 0·58 hr. and 0·60 hr., respectively, for the two methods. The values obtained for the standard deviations of the estimated parameters were much lower, however, for the data fitted by digital computer program, and this would therefore seem to be the better method.

6.4 Curve fitting using analogue computers

The distinction between using analogue computers for simulation and for curve fitting is very small. The procedure is the same as in Chapter 5, except that continuous adjustments are made to the parameter potentiometers until the oscilloscope trace matches the experimental curve. The parameter values are then calculated from the potentiometer settings, the time constants, and the scaling factors. The advantages of this technique are that the computer is easily programmed, the response is rapid so that a wide range of values for a particular parameter can be tested in a very short time, and the parameters estimated are Q_i and k_{ij} directly i.e. the desired parameters do not have to be calculated from complex expressions for X_j and λ_j. The disadvantages are that as the number of parameters to be estimated increases the number of permutations of possible potentiometer settings increases enormously. In addition, it is impossible to calculate even approximate values for the standard deviations of the parameter estimates.

Higinbotham *et al.* (1963) discuss the advantages of displaying the data points and the theoretical curves simultaneously, and describe the computer equipment to do this. Robertson and Cohn (1963) have used this computer to investigate a four compartment model for calcium metabolism (Fig. 6.3). Figure 6.4 shows the oscilloscope trace of the curves fitted to a particular set of experimental data.

6.5 Curve fitting using a digital computer

The use of digital computers in order to improve curve fitting began in the early 1960s. This technique uses a digital computer program to carry out the necessary mathematical steps in minimis-

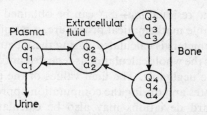

Fig. 6.3 The four compartment model used by Robertson and Cohn (1963) in their analysis of calcium kinetics in man.

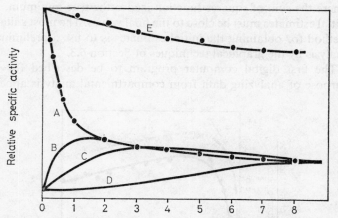

Fig. 6.4 Data points and curves representing relative specific activities of ^{45}Ca in man according to the model of Figure 6.3. Curves A, B, C, and D correspond to a_1, a_2, a_3, and a_4, respectively, and curve E to the total activity $a_1 + a_2 + a_3 + a_4$ (from Robertson and Cohn 1963). (Reproduced with the kind permission of the authors and the New York Academy of Sciences.)

ing the sum of squares function of equation 6.2. The advantages over graphical techniques are that all parameters are determined simultaneously, and are therefore all approximately equally well determined, and that subjective errors or bias on the part of the experimenter are removed.

All procedures require initial estimates for the values of the parameters X_j^0 and λ_j^0. For each experimental point, a_{ik}, t_k, the corresponding value of $a_{ik}^0 = \sum X_j^0 . e^{\lambda_j^0 . t_k}$ is calculated. Hence, the value of the residual $a_{ik} - a_{ik}{}^0$ can be obtained for each point and, by a suitable mathematical procedure, a set of corrections to the initial parameters is calculated. Using these better estimates of the parameters the whole calculation is repeated until the correction factors become negligible. The final values of the parameters are the best estimates and during the computation approximate values for the standard deviations may also be calculated. The most important precaution with this type of iteration procedure is that if the calculation is to converge to a stable set of final parameters, where the sum of squares function is also the true minimum, the initial estimates must be close to the final values. The most suitable method for obtaining the initial estimates is to use a preliminary analysis by the graphical techniques of Section 6.3.

The first digital computer program to be described for the purpose of analyzing data from compartmental analysis appears

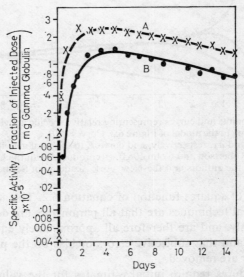

Fig. 6.5 Specific activity of serum ^{131}I-gamma globulin after intrathecal injection (after Lippincott *et al.* 1965). (Reproduced with the kind permission of the editor of the *Journal of Nuclear Medicine*.)

to be that of Worsley and Lax (1962). Its purpose is to fit data to a sum of exponential terms and uses a Gauss–Newton procedure for calculating the parameter corrections. This procedure is described in their paper in more detail than above, and the authors also discuss the problems of convergence and the desirability of using an adequate number of data points. A similar type of program has been described by Richmond *et al.* (1964) and no doubt many other programs for fitting exponentials are now in use.

An example of this type of data analysis is given by Lippincott *et al.* (1965) who investigated the transfer rates of gamma globulin between cerebrospinal fluid and plasma in man. After an intrathecal injection of ^{131}I-gamma globulin, the specific activity of serum gamma globulin was determined giving the results of Figure 6.5. In another series of experiments the specific activity in cerebrospinal fluid (CSF) was determined after an intrathecal injection of labelled gamma globulin with the results shown in Figure 6.6.

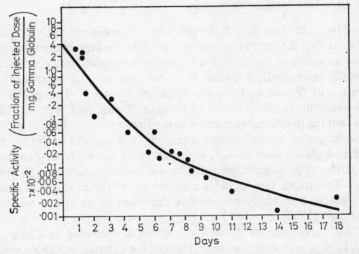

Fig. 6.6 Specific activity of cerebrospinal fluid ^{131}I-gamma globulin after intrathecal injection (after Lippincott *et al.* 1965). (Reproduced with the kind permission of the editor of the *Journal of Nuclear Medicine*.)

The fitted functions for Figure 6.5 were, curve A

$$a = 3{\cdot}19 \times 10^{-5} e^{-0{\cdot}00281\,t} - 3{\cdot}24 \times 10^{-5} e^{-0{\cdot}0366\,t}$$

curve B

$$a = 2{\cdot}12 \times 10^{-5} e^{-0{\cdot}00316\,t} - 2{\cdot}36 \times 10^{-5} e^{-0{\cdot}0251\,t}$$

and for Figure 6.6

$$a = 0{\cdot}0439\,e^{-0{\cdot}0579\,t} + 0{\cdot}00021\,e^{-0{\cdot}00578\,t}$$

where t is in hr. Using the values of X_j and λ_j obtained from these two types of curves the parameters of the system were then calculated. For example,

transfer rate into CSF $= 0{\cdot}425 \pm 0{\cdot}208$ mg/hr.

transfer rate from CSF into plasma $= 1{\cdot}31 \pm 0{\cdot}91$ mg/hr.

rate constant for this transfer $= 0{\cdot}0577$ hr.$^{-1}$

and CSF compartment size $= 22{\cdot}7$ mg

The disadvantage with the program of the type of Worsley and Lax is that the parameters which have been estimated are those of the exponential terms of the expressions obtained after solution of the mathematical model. It is still necessary to calculate the values of Q_i and k_{ij} from the values of X_j and λ_j. This problem is overcome in the program of Berman, Shahn, and Weiss (1962) in that the iteration procedure can use the equations of the original mathematical model, integrate these by a numerical technique to give implicit values of $a_{ik}{}^0$, and then carry out a Gauss–Newton routine to produce corrections to the model parameters Q_i and k_{ij}. This paper also includes a discussion of the problem of convergence. The above program is described in greater detail by Berman (1965).

An application of Berman's program is described by Cohn *et al.* (1965), and their theoretical model for calcium metabolism in man is shown in Figure 6.7. ^{47}Ca was injected intravenously and the specific activities in plasma, urine, and faeces determined.

To use the program the set of differential equations corresponding to each compartment was written out e.g. for plasma,

$$\frac{d(a_1)}{dt} = -(R_{21} + R_{31} + R_{41} + R_{51}) \frac{a_1}{Q_1} + R_{12} \frac{a_2}{Q_2}$$

$$R_{13} \approx 0$$

The data were then fitted to the set of differential equations using Berman's program and the values for the parameters of the model were obtained directly. The results for one experiment in terms of calcium are

$$Q_1 = 2 \cdot 40 \pm 0 \cdot 38 \text{ g}$$
$$Q_2 = 2 \cdot 30 \pm 0 \cdot 37 \text{ g}$$
$$R_{41} = 0 \cdot 072 \pm 0 \cdot 004 \text{ g/day}$$
$$R_{51} = 0 \cdot 230 \pm 0 \cdot 014 \text{ g/day}$$
$$R_{31} = 0 \cdot 356 \pm 0 \cdot 036 \text{ g/day}$$
$$R_{21} = R_{12} = 1 \cdot 89 \pm 0 \cdot 30 \text{ g/day}$$

A second example using this type of program is given in Section 3.12.

In all the programs described so far, the corrections to the parameters are made simultaneously and all the parameters should be equally well determined. Garby et al. (1963) describe a procedure by which some parameters can be better determined than others.

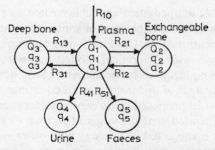

Fig. 6.7 The model used by Cohn et al. (1965) to explain the metabolism of ^{47}Ca in man.

They use the initial estimates for the parameters to calculate a least squares fit for the most critical parameter of the set. Then the other parameters are adjusted using the same procedure, but in some particular sequence determined by the programmer. The whole iteration stage is then repeated until convergence is obtained.

As far as the present author can ascertain, the parameter correction procedures used in all programs published to date employ a Gauss–Newton routine. Many applied scientists are using more recent procedures for parameter estimation based on descent methods for minimising functions. This author uses programs—several examples have been mentioned in this book—based on the papers of Davidon (1959) and Fletcher and Powell (1963), and very often they possess advantages over the more traditional techniques. For example, the difficulties of non-convergence are often reduced (unpublished observations).

In some experimental work the form of the function which is fitted to the experimental data is unknown and many experimenters fit their data to an orthogonal polynomial. This situation rarely occurs in compartmental analysis but Worsley and Lax in their paper (1962) describe a procedure for fitting such polynomials to experimental data using a digital computer program. No applications of this type of program, other than in the above-mentioned paper, appear to have been published.

6.6 Model fitting

A theoretical or mathematical model is created by an experimenter in order to explain the behaviour of some system under study. As previously mentioned in Chapter 2 the technique of compartmental analysis uses theoretical models composed of compartments and the connections between them. The mathematical model is then usually a set of linear differential equations.

The first model proposed by an experimenter is usually based on the current knowledge of the system (his own or other people's), or perhaps it is suggested by analogy with other systems. In some situations a certain amount of intuition may be involved. When it comes to formulating a model to explain a given set of experimental data there is an infinite number of models which do not apply, and probably just as large a number that could fit. For many systems, it may be just as important to be able to discard a model as it is to prove that it is acceptable. When testing the compatibility of the first model with the data, often a preliminary qualitative assessment is made (e.g. by simulation). If this is successful, then a quantitative judgement can be made by deriving the values for the parameters by one of the preceeding techniques. If the model is consistent with the data, then there will be a random scatter of the experimental data about the values predicted by the model. If the scatter is not random, then the model is inconsistent with the data and the former must be modified and retested. The uniqueness of the model is assessed by comparing the standard deviations of the parameter estimates with the parameter estimates themselves. This is a qualitative judgement, but if the ratios are larger than one might expect from the quality of the data, then it often means that several models can equally well be fitted to the data and that the first model is not unique. If a non-unique model is suspected, then either further experiments must be designed to refine the model (e.g. by sampling more compartments), or assumptions and simplifications must be made to the model and the model retested.

In many experiments, a model is not known or proposed for the system in advance. In these cases no model is unique and the usual procedure is to fit the data to the minimum number of exponential terms. Perhaps a quantitative description of the behaviour of the system is all that is required, in which case this procedure would be sufficient. If one wishes to proceed further and formulate a more complex model then the number of exponential terms found by the previous treatment will indicate how many compartments may be required. Also, the values of the parameters may give a suggestion as to how some of the connections are made. One's knowledge of the biochemistry or physiology of the system may place further limitations on the range of the models allowed. The next step is to test the remaining models and eliminate any that are inconsistent with the data. It is unlikely that one unique model is left, so usually one has to continue, as described earlier, in order to improve on the non-unique model or models remaining. The problems of formulating and testing models are discussed further by Berman (1963).

As an example of model fitting we have the experiments of Baker *et al.* (1959). They investigated the quantitative metabolism of glucose in the rat by injecting ^{14}C-glucose and measuring the specific activity of plasma glucose over 300 min. The data were analyzed graphically into three exponentials, and they proposed three possible theoretical models to explain the data (Fig. 6.8). The mathematical models were then formulated, the equations integrated to give a sum of exponentials, and the model parameters calculated using the experimental data. From these results and by comparison with other published data, they chose model C as

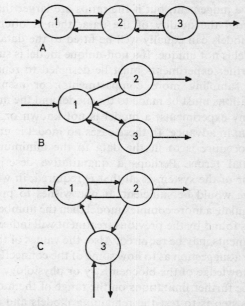

Fig. 6.8 Three possible models to explain the kinetics of ^{14}C-glucose in the rat over a period of about 5 hr. 1: plasma compartment, 2: interstitial fluid, 3: a third compartment.

the most plausible one. Another example of model fitting was given in Section 5.4.

REFERENCES

ATKINS, G. L., DEAN, B. M., GRIFFIN, W. J. and WATTS, R. W. E. (1964). *J. biol. Chem.*, **239**, 2975. Quantitative aspects of ascorbic acid metabolism in man.

BAKER, N., SHIPLEY, R. A., CLARK, R. E. and INCEFY, G. E. (1959). *Am. J. Physiol.*, **196**, 245. C^{14} studies in carbohydrate metabolism: glucose pool size and rate of turnover in the normal rat.

BERMAN, M. (1963). *Ann. N.Y. Acad. Sci.*, **108**, 182. The formulation and testing of models.

BERMAN, M. (1965). Compartmental analysis in kinetics. *In* STACY, R. W. and WAXMAN, B. D. (Eds.) 'Computers in Biomedical Research'. Vol. II, Chapter 7. (Academic Press: London.)

BERMAN, M., SHAHN, E. and WEISS, M. F. (1962). *Biophys. J.*, **2**, 275. The routine fitting of kinetic data to models: a mathematical formalism for digital computers.

COHN, W. E. and BRUES, A. M. (1945). *J. gen. Physiol.*, **28**, 449. Metabolism of tissue cultures. III. A method for measuring the permeability of tissue cells to solutes.

COHN, S. H., BOZZO, S. R., JESSEPH, J. E., CONSTANTINIDES, C., HUENE, D. R. and GUSMANO, E. A. (1965). *Radiat. Res.*, **26**, 319. Formulation and testing of a compartmental model for calcium metabolism in man.

DAVIDON, W. C. (1959). AEC. Research and Development Report, ANL-5990 (Rev). Variable metric method for minimization.

FLETCHER, R. and POWELL, M. J. D. (1963). *Computer J.*, **6**, 163. A rapidly convergent descent method for minimization.

GARBY, L., SCHNEIDER, W., SUNDQUIST, O. and VUILLE, J-C. (1963). *Acta physiol. scand.*, **59**, S. 216, 3. A ferro-erythrokinetic model and its properties.

GLASS, H. I. and DE GARRETA, A. C. (1967). *Physics Med. Biol.*, **12**, 379. Quantitative analysis of exponential curve fitting for biological applications.

HIGINBOTHAM, W. A., SUGARMAN, R. M., POTTER, D. W. and ROBERTSON, J. S. (1963). *Ann. N.Y. Acad. Sci.*, **108**, 117. A direct analog computer for multi-compartment systems.

KRANE, S. M., BROWNELL, G. L., STANBURY, J. B. and CORRIGAN, H. (1956). *J. clin. Invest.*, **35**, 874. The effect of thyroid disease on calcium metabolism in man.

I

LIPPINCOTT, S. W., KORMAN, S., LAX, L. C. and CORCORAN, C. (1965). *J. nucl. Med.*, **6**, 632. Transfer rates of gamma globulin between cerebrospinal fluid and blood plasma (results obtained on a series of multiple sclerosis patients).

MYHILL, J. (1967). *Biophys. J.*, **7**, 903. Investigation of the effect of data error in the analysis of biological tracer data.

RICHMOND, C. R., FURCHNER, J. E., DEAN, P. N. and MCWILLIAMS, P. (1964). *Hlth Phys.*, **10**, 3. Electronic processing and analysis of metabolic data.

ROBERTSON, J. S. and COHN, S. H. (1963). *Ann. N.Y. Acad. Sci.*, **108**, 122. Use of an analog computer in studies of strontium and calcium metabolism in man.

SCHLOERB, P. R., FRÜS-HANSEN, B. J., EDELMAN, I. S., SOLOMON, A. K. and MOORE, F. D. (1950). *J. clin. Invest.*, **29**, 1296. The measurement of total body water in the human subject by deuterium oxide dilution.

VAN LIEW, H. D. (1962). *Science, N.Y.*, **138**, 682. Semilogarithmic plots of data which reflect a continuum of exponential processes.

WORSLEY, B. H. and LAX, L. C. (1962). *Biochim. biophys. Acta*, **59**, 1. Selection of a numerical technique for analyzing experimental data of the decay type with special reference to the use of tracers in biological systems.

WAGNER, J. G. and METZLER, C. M. (1967). *J. pharm. Sci.*, **56**, 658. Estimation of rate constants for absorption and elimination from blood concentration data.

Appendices-Mathematical Methods

In common with all biological investigations of a quantitative nature, compartmental analysis requires a little knowledge of a few mathematical procedures. In these appendices it is intended to produce an elementary account of some of the more useful methods, principally with a view to explaining their application, and no attempt will be made to prove any of the theorems. Such treatments are available in many mathematical textbooks.

A1 The differential operator, D

Operators are often used in mathematics as a convenient notation which leads to the saving of time during the implementation of a mathematical procedure. An operator is defined as a symbol indicating an operation to be performed. The differential operator D represents the first derivative of a function, in compartmental analysis this is usually with respect to time. Thus

$$Dy \equiv \frac{dy}{dt} \equiv y'(t)$$

Also it can be used as follows—

$$Dy \equiv \frac{dy}{dt}; \quad D^2 y \equiv \frac{d^2 y}{dt^2}; \quad D^n y \equiv \frac{d^n y}{dt^n}$$

These elementary operators obey most of the fundamental laws of algebra, so that a set of linear differential equations can be transformed into a set of algebraic equations. For example,

$$\frac{dy_1}{dt} = -k_1 y_2$$

$$\frac{dy_2}{dt} = -k_2 y_1$$

117

becomes

$$Dy_1 + k_1 y_2 = 0$$

$$k_2 y_1 + Dy_2 = 0$$

Treating this as a set of algebraic equations in y_1 and y_2, y_2 can be eliminated

$$(D^2 - k_1 k_2) y_1 = 0$$

or

$$\frac{d^2 y_1}{dt^2} - k_1 k_2 y_1 = 0$$

Hence the differential equation in y_1 can be solved. The method will be explained by applying it to solve the equations of two mathematical models from Chapter 3.

Example 1. Two compartment closed system of Section 3.9. The differential equations describing the concentration of tracer in this system are

$$\frac{d(a_1)}{dt} = k_{12} a_2 \frac{Q_2}{Q_1} - k_{21} a_1 \tag{A1}$$

$$\frac{d(a_2)}{dt} = k_{21} a_1 \frac{Q_1}{Q_2} - k_{12} a_2$$

In operator notation these become

$$Da_1 = k_{12} a_2 \frac{Q_2}{Q_1} - k_{21} a_1$$

$$Da_2 = k_{21} a_1 \frac{Q_1}{Q_2} - k_{12} a_2$$

Rearranging

$$a_1(D + k_{21}) - a_2 k_{12} \frac{Q_1}{Q_2} = 0 \tag{A2}$$

$$a_1 k_{21} \frac{Q_1}{Q_2} - a_2(D + k_{12}) = 0 \tag{A3}$$

Thus our set of differential equations has been transformed into a set of algebraic equations with a_1 and a_2 as the variables. To solve these, multiply equation A2 by $k_{21} Q_1/Q_2$ and equation A3 by $(D+k_{21})$.

$$a_1(D+k_{21})\left(k_{21}\frac{Q_1}{Q_2}\right)-a_2\left(k_{12}\frac{Q_2}{Q_1}\right)\left(k_{21}\frac{Q_1}{Q_2}\right) = 0$$

$$a_1\left(k_{21}\frac{Q_1}{Q_2}\right)(D+k_{21})-a_2(D+k_{12})(D+k_{21}) = 0$$

After subtraction

$$a_2(D+k_{12})(D+k_{21})-a_2 k_{12} k_{21} = 0$$

Or, $$a_2[D+(k_{12}+k_{21})]D = 0 \qquad (A4)$$

Similarly $$a_1[D+(k_{12}+k_{21})]D = 0 \qquad (A5)$$

The general solution to an equation of the form

$$a(D+\lambda_1)(D+\lambda_2)\ldots(D+\lambda_n) = 0$$

is $$a = X_1 e^{-\lambda_1 t}+X_2 e^{-\lambda_2 t}+\cdots+X_n e^{-\lambda_n t}$$

Notice that in equations A4 and A5 one term is of the form $D \equiv (D-0)$ and therefore the solution will have a term in e^0 i.e. $e^0 = 1$. The solution is therefore

$$a_1 = X_1+X_2 . e^{-(k_{12}+k_{21}) t} \qquad (A6)$$

$$a_2 = X_3+X_4 . e^{-(k_{12}+k_{21}) t} \qquad (A7)$$

If the tracer is only added to compartment 1, then the initial conditions of the experiment are that when $t = 0$, $a_1 = a_1(0)$, and $a_2 = 0$. Inserting these initial conditions into equations A6 and A7

$$a_1(0) = X_1 + X_2$$

$$0 = X_3 + X_4$$

Differentiating equation A6, equating this with equation A1 and inserting the initial conditions

$$-k_{21} a_1(0) = -(k_{12} + k_{21}) X_2$$

therefore

$$X_2 = \frac{k_{21} a_1(0)}{(k_{12} + k_{21})}$$

Hence

$$X_1 = \frac{a_1(0) k_{12} + a_1(0) k_{21} - a_1(0) k_{21}}{(k_{12} + k_{21})}$$

$$= \frac{k_{12} a_1(0)}{(k_{12} + k_{21})}$$

By a similar procedure, X_3 and X_4 can be determined

$$k_{21} \frac{Q_1}{Q_2} a_1(0) = -(k_{12} + k_{21}) X_4$$

therefore

$$X_4 = -\frac{k_{21} a_1(0) Q_1}{(k_{12} + k_{21}) Q_2}$$

and

$$X_3 = \frac{k_{21} a_1(0) Q_1}{(k_{12} + k_{21}) Q_2}$$

The solution of the original differential equations is therefore

$$a_1 = \frac{a_1(0)}{(k_{12} + k_{21})} \left[k_{12} + k_{21} e^{-(k_{12} + k_{21}) t} \right]$$

$$a_2 = \frac{k_{21} a_1(0) Q_1}{(k_{12} + k_{21}) Q_2} \left[1 - e^{-(k_{12} + k_{21}) t} \right]$$

Example 2. Three compartment catenary system of Section 3.11.
The equations describing the concentration of tracer in this system
are

$$\frac{d(a_1)}{dt} = -k_{21}a_1$$

$$\frac{d(a_2)}{dt} = k_{21}a_1\frac{Q_1}{Q_1} - k_{32}a_2 \qquad (A8)$$

$$\frac{d(a_3)}{dt} = k_{32}a_2\frac{Q_2}{Q_3} - k_{03}a_3$$

In operator notation they become

$$Da_1 = -k_{21}a_1$$

$$Da_2 = k_{21}a_1\frac{Q_1}{Q_2} - k_{32}a_2$$

$$Da_3 = k_{32}a_2\frac{Q_2}{Q_3} - k_{03}a_3$$

Rearrangement of these equations gives

$$a_1(D+k_{21}) \qquad\qquad\qquad = 0$$

$$a_1\left(\frac{Q_1}{Q_2}k_{21}\right) - a_2(D+k_{32}) \qquad\qquad = 0$$

$$a_2\left(\frac{Q_2}{Q_3}k_{32}\right) - a_3(D+k_{03}) = 0$$

This set of equations can easily be solved to yield

$$a_1(D+k_{21}) = 0$$

$$a_2(D+k_{21})(D+k_{32}) = 0$$

$$a_3(D+k_{21})(D+k_{32})(D+k_{03}) = 0$$

The solution for this new set of equations is

$$a_1 = X_1 e^{-k_{21} t} \tag{A9}$$

$$a_2 = X_2 e^{-k_{21} t} + X_3 e^{-k_{32} t} \tag{A10}$$

$$a_3 = X_4 e^{-k_{21} t} + X_5 e^{-k_{32} t} + X_6 e^{-k_{03} t} \tag{A11}$$

If tracer is only added to compartment 1 then the initial conditions are that when $t = 0$, $a_1 = a_1(0)$ and $a_2 = a_3 = 0$. Inserting these conditions in equations A9, A10, and A11

$$X_1 = a_1(0)$$

$$X_2 + X_3 = 0 \tag{A12}$$

$$X_4 + X_5 + X_6 = 0$$

Differentiating equation A10, equating the result with equation A8 and inserting the initial conditions

$$k_{21} a_1(0) \frac{Q_1}{Q_2} = -k_{21} X_2 - k_{32} X_3 \tag{A13}$$

Equations A12 and A13 are two simultaneous equations in X_2 and X_3. Solving these produces

$$X_2 = \frac{k_{21} a_1(0) Q_1}{(k_{21} - k_{32}) Q_2}$$

and

$$X_3 = \frac{k_{21} a_1(0) Q_1}{(k_{32} - k_{21}) Q_2}$$

By a similar process the solutions for X_4, X_5, and X_6 are

$$X_4 = \frac{k_{32} k_{21} Q_1}{(k_{03} - k_{32})(k_{32} - k_{21}) Q_3}$$

$$X_5 = \frac{k_{32} k_{21} Q_1}{(k_{21} - k_{03})(k_{32} - k_{21}) Q_3}$$

$$X_6 = \frac{k_{32} k_{21} Q_1}{(k_{32} - k_{03})(k_{21} - k_{03}) Q_3}$$

Hence the solution to the original set of differential equations has been determined.

A2 The Laplace transform

The Laplace transform is defined as follows. If a function of time, $f(t)$, is defined for all values of t greater than 0, and if s is a real number such that the integral $F(s)$, defined by

$$F(s) = \int_0^\infty e^{-st} f(t)\, dt \qquad (A14)$$

converges for some finite value of s and all greater values, then $F(s)$ is the Laplace transform of $f(t)$. This transformation is useful because a linear differential equation can be transformed into a linear algebraic equation. The latter can then be readily solved. The procedure can be compared to using logarithms where the process of multiplication is converted into one of addition. In practice there is no need to integrate equation A14 each time the Laplace transformation is used, the transforms of the commonest functions can be obtained from tables. The functions frequently used in compartmental analysis are given in Table A1. After the algebraic equations have been solved, the result is again transformed to obtain the solution to the original differential

TABLE A1

Laplace transforms of functions commonly used in compartmental analysis.

Function	Transform
$f(t)$	$F(s)$
$a . f_1(t) + b . f_2(t)$	$a . F_1(s) + b . F_2(s)$
$\dfrac{df(t)}{dt}$	$s . F(s) - f(0)$
a	$\dfrac{a}{s}$
$b . e^{-at}$	$\dfrac{b}{(s+a)}$

K

equations. To illustrate the use of the Laplace transform two examples will be taken from Chapter 3.

Example 3. Two compartment catenary system of Section 3.5. The equations for the concentration of tracer in the two compartments are

$$\frac{d(a_1)}{dt} = -k_{21} a_1 \tag{A15}$$

$$\frac{d(a_2)}{dt} = k_{21} a_1 \frac{Q_1}{Q_2} - k_{02} a_2 \tag{A16}$$

Let $a_1 = f_1(t)$ and $a_2 = f_2(t)$. Substitution in equations A15 and A16 gives

$$\frac{df_1(t)}{dt} = -k_{21} f_1(t) \tag{A17}$$

$$\frac{df_2(t)}{dt} = k_{21} \frac{Q_1}{Q_2} f_1(t) - k_{01} f_2(t) \tag{A18}$$

Using the Laplace transforms for $df(t)/dt$ and $f(t)$, equations A17 and A18 become

$$sF_1(s) - f_1(0) = -k_{21} F_1(s) \tag{A19}$$

$$sF_2(s) - f_2(0) = k_{21} \frac{Q_1}{Q_2} F_1(s) - k_{02} F_2(s) \tag{A20}$$

Hence the differential equations are now transformed into a set of linear algebraic equations with $F_1(s)$ and $F_2(s)$ as the variables. If tracer is only injected into compartment 1, then the initial conditions are $f_1(0) = a_1(0)$ and $f_2(0) = 0$. Inserting these into equations A19 and A20

$$sF_1(s) - a_1(0) = -k_{21} F_1(s)$$

$$sF_2(s) = k_{21} \frac{Q_1}{Q_2} F_1(s) - k_{02} F_2(s)$$

Rearranging these equations gives

$$F_1(s)(s+k_{21}) = a_1(0)$$

$$F_1(s)k_{21}\frac{Q_1}{Q_2} - F_2(s)(s+k_{02}) = 0$$

Their solution is

$$F_1(s) = \frac{a_1(0)}{(s+k_{21})}$$

$$F_2(s) = \frac{k_{21}a_1(0)\dfrac{Q_1}{Q_2}}{(s+k_{21})(s+k_{02})} \tag{A21}$$

Let x and y be two constants so that

$$F_2(s) = \frac{x}{(s+k_{21})} + \frac{y}{(s+k_{02})}$$

$$= \frac{s(x+y)+(xk_{02}+yk_{21})}{(s+k_{21})(s+k_{02})}$$

Comparison with equation A21 shows that

$$x+y = 0 \tag{A22}$$

$$k_{02}x+k_{21}y = k_{21}a_1(0)\frac{Q_1}{Q_2} \tag{A23}$$

Solution of equations A22 and A23 yields

$$y = -x = \frac{k_{21}a_1(0)Q_1}{(k_{21}-k_{02})Q_2}$$

so that equation A21 expressed as partial fractions becomes

$$F_2(s) = \frac{k_{21}a_1(0)Q_1}{(k_{21}-k_{02})Q_2}\cdot\frac{1}{(s+k_{02})} - \frac{k_{21}a_1(0)Q_1}{(k_{21}-k_{02})Q_2}\cdot\frac{1}{(s+k_{21})}$$

Using the inverse Laplace transform for $b/(s+a)$, the solution to the original set of differential equations is

$$a_1 = a_1(0) \cdot e^{-k_{21} t}$$

and
$$a_2 = \frac{k_{21} a_1(0) Q_1}{(k_{21} - k_{02}) Q_2} \left[e^{-k_{02}t} - e^{-k_{21}t} \right]$$

Example 4. The *two compartment system* of Section 3.10. The concentration of tracer in this system is described by

$$\frac{d(a_1)}{dt} = k_{12} a_2 \frac{Q_2}{Q_1} - k_{21} a_1 - k_{01} a_1 \qquad (A24)$$

$$\frac{d(a_2)}{dt} = k_{21} a_1 \frac{Q_1}{Q_2} - k_{12} a_2 \qquad (A25)$$

Substituting $a_1 = f_1(t)$ and $a_2 = f_2(t)$, then equations A24 and A25 become

$$\frac{df_1(t)}{dt} = k_{12} \frac{Q_2}{Q_1} f_2(t) - k_{21} f_1(t) - k_{01} f_1(t)$$

$$\frac{df_2(t)}{dt} = k_{21} \frac{Q_1}{Q_2} f_1(t) - k_{12} f_2(t)$$

Using the Laplace transforms for $df(t)/dt$ and $f(t)$, then two algebraic equations are obtained

$$sF_1(s) - f_1(0) = k_{12} \frac{Q_2}{Q_1} F_2(s) - k_{21} F_1(s) - k_{01} F_1(s) \quad (A26)$$

$$sF_2(s) - f_2(0) = k_{21} \frac{Q_1}{Q_2} F_1(s) - k_{12} F_2(s) \qquad (A27)$$

If the tracer is only placed into compartment 1, then the initial conditions are $f_1(0) = a_1(0)$ and $f_2(0) = 0$. Using these in equations A26 and A27

$$sF_1(s) - a_1(0) = k_{12} \frac{Q_2}{Q_1} F_2(s) - k_{21} F_1(s) - k_{01} F_1(s)$$

$$sF_2(s) = k_{21} \frac{Q_1}{Q_2} F_1(s) - k_{12} F_2(s)$$

Rearranging

$$(s+k_{21}+k_{01})F_1(s)-\left(k_{12}\frac{Q_2}{Q_1}\right)F_2(s) = a_1(0)$$

$$-\left(k_{21}\frac{Q_1}{Q_2}\right)F_1(s)+(s+k_{12})F_2(s) = 0$$

Solution of these equations gives

$$F_1(s) = \frac{a_1(0)(s+k_{12})}{(s+k_{21}+k_{01})(s+k_{12})-k_{21}k_{12}}$$

and

$$F_2(s) = \frac{a_1(0)k_{21}\dfrac{Q_1}{Q_2}}{(s+k_{21}+k_{01})(s+k_{12})-k_{21}k_{12}}$$

These equations can be resolved into partial fractions, but the steps are also left to the reader. The result of this operation is

$$F_1(s) = \frac{a_1(0)(\lambda_1-k_{12})}{(\lambda_1-\lambda_2)(s+\lambda_1)}+\frac{a_1(0)(k_{12}-\lambda_2)}{(\lambda_1-\lambda_2)(s+\lambda_2)}$$

$$F_2(s) = \frac{a_1(0)k_{21}\dfrac{Q_1}{Q_2}}{(\lambda_1-\lambda_2)(s+\lambda_1)}-\frac{a_1(0)k_{21}\dfrac{Q_1}{Q_2}}{(\lambda_1-\lambda_2)(s+\lambda_2)}$$

where

$$\lambda_1 = \frac{-(k_{12}+k_{21}+k_{01})-\sqrt{(k_{12}+k_{21}+k_{01})^2-4k_{01}k_{12}}}{2}$$

$$\lambda_2 = \frac{-(k_{12}+k_{21}+k_{01})-\sqrt{(k_{12}+k_{21}+k_{01})^2-4k_{01}k_{12}}}{2}$$

The solution to the original set of differential equations is obtained by performing the inverse Laplace transformation

$$a_1 = \frac{a_1(0)}{(\lambda_1-\lambda_2)}\left[(\lambda_1-k_{12})e^{-\lambda_1 t}+(k_{12}-\lambda_2)e^{-\lambda_2 t}\right]$$

$$a_2 = \frac{a_1(0)k_{21}\dfrac{Q_1}{Q_2}}{(\lambda_1-\lambda_2)}\left[e^{-\lambda_1 t}-e^{-\lambda_2 t}\right]$$

A3 Matrices and determinants

Matrices are used in Section A4 as a convenient means of handling a large number of linear algebraic equations. In this Section, the algebra of matrices is briefly outlined and many of the examples used are taken from Section A4.

A matrix of order $m \times n$ is a rectangular array of numbers arranged in m rows and n columns. The matrix $\mathbf{A}$ is defined by

$$\mathbf{A} = \begin{bmatrix} a_{11} & a_{12} & \cdots & a_{1n} \\ a_{21} & a_{22} & \cdots & a_{2n} \\ \cdots\cdots\cdots\cdots\cdots\cdots \\ a_{m1} & a_{m2} & \cdots & a_{mn} \end{bmatrix}$$

The symbol $\mathbf{A}$ represents a matrix and the symbol a_{ij} represents an element of the matrix in the ith row and jth column, where $i = 1, 2, \ldots m$, and $j = 1, 2, \ldots n$.

If $m = n$ then $\mathbf{A}$ is a square matrix of order n. If $n = 1$ then $\mathbf{A}$ is a column matrix of order $m \times 1$, and if $m = 1$ then $\mathbf{A}$ is a row matrix of order $1 \times n$.

Thus

$$\begin{bmatrix} X_{11} & X_{12} & \cdots & X_{1n} \\ X_{21} & X_{22} & \cdots & X_{2n} \\ \cdots\cdots\cdots\cdots\cdots\cdots \\ X_{n1} & X_{n2} & \cdots & X_{nn} \end{bmatrix}$$ is a square matrix,

$$\begin{bmatrix} A_1 \\ A_2 \\ . \\ . \\ . \\ A_n \end{bmatrix}$$ is a column matrix,

and $\begin{bmatrix} Y_1 & Y_2 & \cdots & Y_n \end{bmatrix}$ a row matrix.

Two matrices may be *added* or *subtracted* if they are of the same order. If the elements of the two matrices $\mathbf{B}$ and $\mathbf{A}$ are a_{ij} and b_{ij}, then the elements of $\mathbf{C} = \mathbf{A} \pm \mathbf{B}$ are $a_{ij} \pm b_{ij}$. An example from Section A4 is

$$\mathbf{sI} - \mathbf{k} =$$

$$\begin{bmatrix} s & 0 & \ldots & 0 \\ 0 & s & \ldots & 0 \\ & \ldots\ldots & & \\ 0 & 0 & \ldots & s \end{bmatrix} - \begin{bmatrix} -k_{11} & k_{12} & \ldots & k_{1n} \\ k_{21} & -k_{22} & \ldots & k_{2n} \\ & \ldots\ldots\ldots\ldots & & \\ k_{n1} & k_{n2} & \ldots & k_{nn} \end{bmatrix} = \begin{bmatrix} (s+k_{11}) & -k_{12} & \ldots -k_{1n} \\ -k_{21} & (s+k_{22}) & \ldots -k_{2n} \\ & \ldots\ldots\ldots\ldots & \\ -k_{n1} & -k_{n2} & \ldots (s+k_{nn}) \end{bmatrix}$$

Multiplication by a constant is the same as repetitive addition. If $\mathbf{B} = p.\mathbf{A}$, then the elements of $\mathbf{B}$ are $p.a_{ij}$. Thus the matrix used above $\mathbf{sI} = s.\mathbf{I}$

$$s \times \begin{bmatrix} 1 & 0 \ldots 0 \\ 0 & 1 \ldots 0 \\ & \ldots\ldots \\ 0 & 0 \ldots 1 \end{bmatrix} = \begin{bmatrix} s & 0 \ldots 0 \\ 0 & s \ldots 0 \\ & \ldots\ldots \\ 0 & 0 \ldots s \end{bmatrix}$$

Multiplication of two matrices $\mathbf{A}$ and $\mathbf{B}$ can be performed provided the number of columns in $\mathbf{A}$ equals the number of rows in $\mathbf{B}$. That is $\mathbf{A}$ must be of order $r \times n$ and $\mathbf{B}$ $n \times s$. The product $\mathbf{C} = \mathbf{A}.\mathbf{B}$ has order $r \times s$, and its elements are given by

$$C_{ij} = \sum a_{ip}.b_{pj} \quad (p = 1, 2, \ldots n)$$

Thus

$$\begin{bmatrix} a & b & c \\ d & e & f \end{bmatrix} \times \begin{bmatrix} u & v \\ w & x \\ y & z \end{bmatrix} = \begin{bmatrix} (au+bw+cy) & (av+bx+cz) \\ (du+ew+fy) & (dv+ex+fz) \end{bmatrix}$$

$$2 \times 3 \qquad\qquad 3 \times 2 \qquad\qquad\qquad 2 \times 2$$

If $\mathbf{A}$ is multiplied by $\mathbf{B}$ thus, $\mathbf{C} = \mathbf{B}.\mathbf{A}$, then $\mathbf{A}$ is premultiplied by $\mathbf{B}$. If the product is obtained by $\mathbf{C} = \mathbf{A}.\mathbf{B}$ then this is post-multiplication. Note that the two results are quite different,

$$\begin{bmatrix} 1 & 2 \\ 4 & 3 \end{bmatrix} \times \begin{bmatrix} 2 & 3 \\ 5 & 2 \end{bmatrix} = \begin{bmatrix} 12 & 7 \\ 23 & 8 \end{bmatrix}$$

$$\begin{bmatrix} 2 & 3 \\ 5 & 2 \end{bmatrix} \times \begin{bmatrix} 1 & 2 \\ 4 & 3 \end{bmatrix} = \begin{bmatrix} 14 & 13 \\ 13 & 16 \end{bmatrix}$$

A *diagonal matrix* is a square matrix in which all the elements except those on the diagonal are zero. Thus $\mathbf{A}$ is a diagonal matrix,

$$\mathbf{A} = \begin{bmatrix} a_{11} & 0 & \dots & 0 \\ 0 & a_{22} & \dots & 0 \\ \multicolumn{4}{c}{\dotfill} \\ 0 & 0 & \dots & a_{nn} \end{bmatrix} = \begin{bmatrix} a_{11} & & & 0 \\ & a_{22} & & \\ & & \cdot & \\ 0 & & & a_{nn} \end{bmatrix}$$

The inverse (see later) of a diagonal matrix is important and has the following useful property.

$$\text{If } \mathbf{A} = \begin{bmatrix} a_1 & & & & \\ & a_2 & & 0 & \\ & & a_3 & & \\ & 0 & & \cdot & \\ & & & & a_n \end{bmatrix} \text{ then } \mathbf{A}^{-1} = \begin{bmatrix} \dfrac{1}{a_1} & & & & \\ & \dfrac{1}{a_2} & & 0 & \\ & & \dfrac{1}{a_3} & & \\ & 0 & & \cdot & \\ & & & & \dfrac{1}{a_n} \end{bmatrix}$$

Premultiplying a matrix by a diagonal matrix produces the result,

$$\begin{bmatrix} u & & & \\ & v & & 0 \\ & & \cdot & \\ & 0 & & \cdot \\ & & & z \end{bmatrix} \times \begin{bmatrix} a_{11} & a_{12} \dots a_{1n} \\ a_{21} & a_{22} \dots a_{2n} \\ \multicolumn{2}{c}{\dotfill} \\ a_{n1} & a_{n2} \dots a_{nn} \end{bmatrix} = \begin{bmatrix} ua_{11} & ua_{12} \dots ua_{1n} \\ va_{21} & va_{22} \dots va_{2n} \\ \multicolumn{2}{c}{\dotfill} \\ za_{n1} & za_{n2} \dots za_{nn} \end{bmatrix}$$

whereas postmultiplying gives

$$\begin{bmatrix} a_{11} & a_{12} \dots a_{1n} \\ a_{21} & a_{22} \dots a_{2n} \\ \multicolumn{2}{c}{\dotfill} \\ a_{n1} & a_{n2} \dots a_{nn} \end{bmatrix} \times \begin{bmatrix} u & & & \\ & v & & 0 \\ & & \cdot & \\ & 0 & & \cdot \\ & & & z \end{bmatrix} = \begin{bmatrix} ua_{11} & va_{12} \dots za_{1n} \\ ua_{21} & va_{22} \dots za_{2n} \\ \multicolumn{2}{c}{\dotfill} \\ ua_{n1} & va_{n2} \dots za_{nn} \end{bmatrix}$$

In the above two examples if all the elements of the diagonal are equal i.e. if $u = v = \cdots = z$, this is equivalent to multiplying the matrix $\mathbf{A}$ by a constant. A special case of this type of matrix is the unit matrix $\mathbf{I}$ where the diagonal elements are 1. Thus,

$$\mathbf{I} = \begin{bmatrix} 1 & & & \\ & 1 & & 0 \\ & & \cdot & \\ & 0 & & \cdot \\ & & & & 1 \end{bmatrix} = 1$$

The *transpose* of the $n \times m$ matrix $\mathbf{A}$, written as $\mathbf{A}^T$ or $\mathbf{A}'$, is an $m \times n$ matrix obtained by interchanging the rows of $\mathbf{A}$. For example

$$\text{if } \mathbf{A} = \begin{bmatrix} a & b & c \\ d & e & f \end{bmatrix} \quad \mathbf{A}^T = \begin{bmatrix} a & d \\ b & e \\ c & f \end{bmatrix}$$

A *determinant* is the property of a square matrix and is written thus, $|\mathbf{A}|$ as the determinant of $\mathbf{A}$. Whereas a matrix is an operator, a determinant is a real number, and is calculated as follows for a 3×3 matrix.

$$\text{If } \mathbf{A} = \begin{bmatrix} a_{11} & a_{12} & a_{13} \\ a_{21} & a_{22} & a_{23} \\ a_{31} & a_{32} & a_{33} \end{bmatrix} \quad \text{then}$$

$$|\mathbf{A}| = (a_{11}a_{22}a_{33}) - (a_{11}a_{23}a_{32}) - (a_{12}a_{21}a_{33}) + (a_{12}a_{23}a_{31})$$
$$+ (a_{13}a_{21}a_{32}) - (a_{13}a_{22}a_{31})$$

For an $n \times n$ matrix there are $n!$ products in its determinant, and each product is of the form $(a_{1i}a_{2j} \ldots a_{nk})$ where $i \neq j \neq \cdots \neq k$. There are several rules for determining the sign of the product, one such rule is as follows. The elements in the product are ranked in order of row indices, as in the example above. The column indices are then written down and the number of 'inversions' present is calculated. Thus 123—no inversions, 213 one inversion of the numbers 1 and 2. The complete list for the 3×3 matrix is

TABLE A2

Complete list of inversions for a 3×3 matrix.

Column indices	No. of inversions	Sign
123	0	+
132	1	−
213	1	−
231	2	+
312	2	+
321	3	−

given in Table A2. If the number of inversions present is even then the sign of the product is positive, if the number is odd then the sign is negative. The value of the determinant is zero if (1) all the elements in any row or column are zero, (2) if two rows or two columns are equal. This is important during the inversion of matrices.

The *minor* of the element a_{ij} of the square matrix $\mathbf{A}$ is the determinant $|\mathbf{M}_{ij}|$ obtained by the elimination of the ith row and jth column of the determinant $|\mathbf{A}|$.

$$\text{Thus if } \mathbf{A} = \begin{bmatrix} a_{11} & a_{12} & \cdots & a_{1n} \\ a_{21} & a_{22} & \cdots & a_{2n} \\ a_{31} & a_{32} & \cdots & a_{3n} \\ \cdots\cdots\cdots\cdots\cdots \\ a_{n1} & a_{n2} & \cdots & a_{nn} \end{bmatrix}$$

$$\text{then } |\mathbf{M}_{12}| = \begin{bmatrix} a_{21} & a_{23} & \cdots & a_{2n} \\ a_{31} & a_{33} & \cdots & a_{3n} \\ \cdots\cdots\cdots\cdots\cdots \\ a_{n1} & a_{n2} & \cdots & a_{nn} \end{bmatrix}$$

The *cofactor* of the element a_{ij}, written $\mathbf{C}_{ij}$, is the minor $|\mathbf{M}_{ij}|$ multiplied by an appropriate sign, thus

$$\mathbf{C}_{ij} = (-1)^{i+j} \cdot |\mathbf{M}_{ij}|$$

The *inverse matrix* of the square matrix $\mathbf{A}$, written $\mathbf{A}^{-1}$, is defined by

$$\mathbf{A}^{-1} = \frac{\mathbf{C}_{ij}{}^{T}}{|\mathbf{A}|} \tag{A28}$$

The matrix C_{ij}^T is the transpose of a matrix whose elements are the cofactors of A. Note that if $|A| = 0$ then A cannot be inverted. The inverse matrix has the important property that

$$AA^{-1} = A^{-1}A = I = 1$$

A matrix cannot be divided by another matrix. Instead one must multiply by the inverse of the matrix. Thus if $BC = A$, then

$$C = \frac{A}{B} = B^{-1}A$$

and

$$B = \frac{A}{C} = AC^{-1}$$

Use of matrices in compartmental analysis. If there are n simultaneous algebraic equations containing n unknown values of x,

$$\left.\begin{array}{l} a_{11}x_1 + a_{12}x_2 + \cdots + a_{1n}x_n = b_1 \\ \dotfill \\ a_{n1}x_1 + a_{n2}x_2 + \cdots + a_{nn}x_n = b_n \end{array}\right\} \quad (A29)$$

then they may be written concisely in matrix form as

$$A . x = b \qquad (A30)$$

A is the $n \times n$ matrix
$$\begin{bmatrix} a_{11} & \cdots & a_{1n} \\ \dotfill \\ a_{n1} & \cdots & a_{nn} \end{bmatrix}$$

and x and b are the column matrices
$$\begin{bmatrix} x_1 \\ \cdot \\ \cdot \\ \cdot \\ x_n \end{bmatrix} \quad \text{and} \quad \begin{bmatrix} b_1 \\ \cdot \\ \cdot \\ \cdot \\ b_n \end{bmatrix}$$

The equivalence of equations A29 and A30 can be demonstrated by multiplying out Ax. Premultiplying equation A30 by A^{-1} gives

$$x = A^{-1}b$$

Alternatively, using equation A28

$$\mathbf{x} = \frac{\mathbf{C}_{ij}{}^T \mathbf{b}}{|\mathbf{A}|}$$

Hence the values of $\mathbf{x}$ can be calculated provided $|\mathbf{A}|$ is not equal to, or very close to, zero.

In Section A4 sets of simultaneous differential equations are converted to linear algebraic equations by the Laplace transformation. Matrix algebra is then used in the solution of these equations, and hence the original differential equations can be solved.

A4 Kinetics of systems in a steady state—general systems and complex systems

A4.1 Introduction

The treatments in Chapter 3 became more and more complex as a system increased in size. The labour involved in solving the mathematical equations for a system can be much reduced if a solution is available for a general n-compartment system. The solution for a particular system can then be obtained by using suitable substitutions.

Several general treatments have been proposed, particularly by Sheppard (1948), Sheppard and Householder (1951), and Berman and Schoenfeld (1956). The procedure outlined in the third paper is the most elegant and has found wide interest. The same procedure had, however, been outlined by Tobias (1949), but the solution for a general system now described will be based largely on the paper of Berman and Schoenfeld. Closed systems will be discussed first because the treatment is simplest, and then adaptations to some other systems will be described.

A4.2 Closed systems

In a general n-compartment closed system each compartment is connected to every other compartment in the system, and there are no connections outside the system. Such a system can be represented by the theoretical model shown in Figure A1. Taking each compartment in turn, we can write the first order differential equations describing the rate of change of concentration for the traced substance, Q_j, with time. These equations form the mathematical model.

$$\left.\begin{aligned}
\frac{d(Q_1)}{dt} &= k_{12}Q_2 + k_{13}Q_3 + \cdots + k_{1n}Q_n - k_{21}Q_1 - k_{31}Q_1 - \cdots - k_{n1}Q_1 \\
&\cdots\cdots\cdots\cdots\cdots\cdots\cdots\cdots\cdots\cdots\cdots\cdots\cdots\cdots\cdots\cdots\cdots\cdots\cdots\cdots\cdots\cdots\cdots\cdots \\
\frac{d(Q_n)}{dt} &= k_{n1}Q_1 + k_{n2}Q_2 + \cdots + k_{n,n-1}Q_{n-1} - k_{1n}Q_n - k_{2n}Q_n - \cdots - k_{n-1,n}Q_n
\end{aligned}\right\} \quad (A31)$$

Because the transport of substance from the jth compartment to itself is meaningless, the terms k_{jj} are omitted. It is more convenient and concise if the above equations are expressed in summation notation, so that equation A31 becomes

$$\frac{d(Q_i)}{dt} = \sum k_{ij}Q_j - Q_i \sum k_{ji} \quad (j = 1, 2, \ldots n; j \neq i) \quad (A32)$$

Because the system is in a steady state, the rate at which substance enters a compartment equals the rate at which it leaves, so that

$$\frac{d(Q_i)}{dt} = 0 \quad (A33)$$

The concentration of the tracer can be expressed by a similar set of equations to equation A32, thus

$$\frac{d(q_i)}{dt} = \sum k_{ij}q_j - q_i \sum k_{ji} \quad (A34)$$

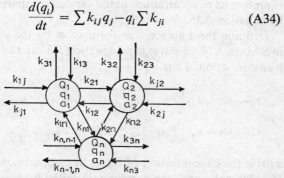

Fig. A1 A general n-compartment system in which each compartment is connected to every other compartment in the system.

We have already noted that $d(q_i)/dt$ can be replaced by $[d(a_i)/dt]\,Q_i$, hence equation A34 becomes

$$\frac{d(a_i)}{dt} = \frac{1}{Q_i}\sum k_{ij}\,a_j\,Q_j - a_i\sum k_{ji} \qquad (A35)$$

Because the tracer is introduced into only one compartment, its concentration in each compartment does not reach a steady state until t approaches infinity. Therefore

$$\frac{d(a_i)}{dt} \neq 0$$

Since we are only concerned with closed systems for the moment, the total amount of tracer in the system is constant, so that

$$\sum \frac{d(q_i)}{dt} = 0 \quad (j = 1, 2, \ldots n) \qquad (A36)$$

It can be noted that in a n-compartment system there are n equations but because of the relationships equation A33 or equation A36 only $(n-1)$ of these equations are independent. The mathematical model for the general n-compartment closed system can therefore be expressed in the form of equation A32. When tracer is present its concentrations in the various compartments are given by equation A35.

Defining the Laplace transform of a_i by the symbol A_i ($F_i(s)$ in Section A2), we can expand equation A35 and apply the Laplace transformation. Thus,

$$sA_1 - a_1(0) = k_{12}\frac{Q_2}{Q_1}A_2 + k_{13}\frac{Q_3}{Q_1}A_3 + \cdots + k_{1n}\frac{Q_n}{Q_1}A_n - A_1\sum k_{j1}$$

$$sA_n - a_n(0) = k_{n1}\frac{Q_1}{Q_n}A_1 + k_{n2}\frac{Q_2}{Q_n}A_2 + \cdots + k_{n,n-1}\frac{Q_{n-1}}{Q_n}A_{n-1} - A_n\sum k_{jn}$$

$a_i(0)$ is the concentration of tracer in compartment i when $t = 0$. Usually only one compartment is initially labelled so that all except one value of $a_i(0)$ is zero. Rearranging gives

$$(s+\sum k_{j1})A_1 - \quad k_{12}\frac{Q_2}{Q_1}A_2 - \cdots - \quad k_{1n}\frac{Q_n}{Q_1}A_n = a_1(0)$$

$$-k_{21}\frac{Q_1}{Q_2}A_1 + (s+\sum k_{j2})A_2 - \cdots - \quad k_{2n}\frac{Q_n}{Q_2}A_n = a_2(0)$$

$$\cdots\cdots\cdots\cdots\cdots\cdots\cdots\cdots\cdots\cdots\cdots\cdots\cdots\cdots\cdots\cdots\cdots\cdots\cdots\cdots\cdots\cdots$$

$$-k_{n1}\frac{Q_1}{Q_n}A_1 - \quad k_{n2}\frac{Q_2}{Q_n}A_2 - \cdots + (s+\sum k_{jn})A_n = a_n(0)$$

$$\text{(A37)}$$

The solution of these equations will be carried out using matrices. Section A3 gives a brief outline of some of the principles needed for this solution. The following matrices are defined

I is the unit matrix
$$\begin{bmatrix} 1 & & & & & \\ & 1 & & 0 & & \\ & & \cdot & & & \\ & 0 & & \cdot & & \\ & & & & & \\ & & & & & 1 \end{bmatrix}$$

A and a(0) are column matrices
$$\begin{bmatrix} A_1 \\ A_2 \\ \cdot \\ \cdot \\ \cdot \\ A_n \end{bmatrix} \quad \begin{bmatrix} a_1(0) \\ a_2(0) \\ \cdot \\ \cdot \\ \cdot \\ a_n(0) \end{bmatrix}$$

Q is the diagonal matrix
$$\begin{bmatrix} Q_1 & & & & \\ & Q_2 & & 0 & \\ & & \cdot & & \\ & 0 & & \cdot & \\ & & & & Q_n \end{bmatrix}$$

k is the square matrix
$$\begin{bmatrix} -k_{11} & k_{12} & \dots & k_{1n} \\ k_{21} & -k_{22} & \dots & k_{2n} \\ \dots\dots\dots\dots\dots\dots\dots \\ k_{n1} & k_{n2} & \dots & -k_{nn} \end{bmatrix}$$

where the diagonal elements k_{ii} are negative. Since these elements have no significance in equations A32, A35, or A37 we can put

$$k_{ii} = \sum k_{ji} \quad (j = 1, 2, \dots n; j \neq i)$$

Then the equations A37 in matrix form become

$$\mathbf{Q}^{-1}(s\mathbf{I} - \mathbf{k})\,\mathbf{Q}\mathbf{A} = \mathbf{a}(0)$$

Rearranging in steps,

$$(s\mathbf{I} - \mathbf{k})\,\mathbf{Q}\mathbf{A} = \mathbf{Q}\mathbf{a}(0)$$

$$\mathbf{Q}\mathbf{A} = (s\mathbf{I} - \mathbf{k})^{-1}\,\mathbf{Q}\mathbf{a}(0)$$

$$\mathbf{A} = \mathbf{Q}^{-1}(s\mathbf{I} - \mathbf{k})^{-1}\,\mathbf{Q}\mathbf{a}(0) \qquad (A38)$$

Let Δ be the determinant of $(s\mathbf{I} - \mathbf{k})$

Δ_{ij} be the ith row and jth column cofactor of $(s\mathbf{I} - \mathbf{k})$

$\boldsymbol{\Delta}_{ij}$ be the matrix of cofactors of $(s\mathbf{I} - \mathbf{k})$

and $\boldsymbol{\Delta}_{ij}^{T}$ be the transpose of $\boldsymbol{\Delta}_{ij}$

Then (see Section A3), $\quad (s\mathbf{I} - \mathbf{k})^{-1} = \dfrac{\boldsymbol{\Delta}_{ij}^{T}}{\Delta}$ (A39)

Substitution of equation A39 into equation A38 produces

$$A_j = \sum \frac{\Delta_{ij}}{\Delta}\frac{Q_i}{Q_j}\,a_i(0) \qquad (A40)$$

The determinant Δ is a polynomial in s of nth degree. In a steady state system, because $k_{ij} \geqslant 0$, the roots of the equation $\Delta = 0$ are real and negative. Imaginary roots are impossible (Hearon

1953; Berman and Schoenfeld 1956.) Let these n roots be $-\lambda_j$ $(j = 1, 2, \ldots n)$, noting that for a closed system one root exists at $s = 0$ (Berman and Schoenfeld 1956). It has been shown that the order of magnitude of the roots is $-\lambda_1 > -\lambda_2 > \cdots > -\lambda_n$ (Sheppard and Householder 1951), therefore in the closed system $-\lambda_1$ is the root equal to zero. The terms

$$\sum \frac{\Delta_{ij}}{\Delta} \frac{Q_i}{Q_j} a_i(0)$$

of equation A40 can thus be resolved into a series of terms of the form $X_i/(s+\lambda_i)$, so that equation A40 becomes

$$A_j = \sum \frac{X_i}{(s+\lambda_i)} \quad (i = 1, 2, \ldots n) \tag{A41}$$

The case of multiple roots (i.e. $\lambda_p = \lambda_q$) rarely occurs in compartmental analysis, because the roots cannot be separated during curve analysis (see Chapter 6).

The inverse Laplace transform of equation A41 yields

$$a_j = \sum X_{ji} \cdot e^{-\lambda_i t} \quad (i = 1, 2, \ldots n) \tag{A42}$$

The values of $\mathbf{X}$ and $\boldsymbol{\lambda}$ are derived from the experimental data by one of the curve fitting methods of Chapter 6. These are termed invariants because they depend on the parameters of the system and not on the results obtained from an experiment. Substitution of equation A42 into equation A35 produces the set of n^2 equations

$$X_{1i}\sum k_{j1} - \frac{Q_2}{Q_1}k_{12}X_{2i} - \frac{Q_3}{Q_1}k_{13}X_{3i} - \cdots - \frac{Q_n}{Q_1^0}k_{1n}X_{ni} = \lambda_i X_{1i}$$

$$\cdots\cdots\cdots\cdots\cdots\cdots\cdots\cdots\cdots\cdots\cdots\cdots\cdots\cdots$$

$$-\frac{Q_1}{Q_n}k_{n1}X_{1i} - \frac{Q_2}{Q_n}k_{n2}X_{2i} - \frac{Q_3}{Q_n}k_{n3}X_{3i} - \cdots + X_{ni}\sum k_{jn} = \lambda_i X_{ni}$$

$$(j = 1, 2, \ldots n; \quad i = 1, 2, \ldots n; i \neq j) \tag{A43}$$

L

If the following matrices are now defined

λ is the diagonal
matrix

$$\begin{bmatrix} -\lambda_1 & & & & \\ & -\lambda_2 & & & 0 \\ & & \cdot & & \\ 0 & & & \cdot & \\ & & & & -\lambda_n \end{bmatrix}$$

X is the square matrix

$$\begin{bmatrix} X_{11} & \cdots & X_{1n} \\ \cdots\cdots\cdots\cdots \\ X_{n1} & \cdots & X_{nn} \end{bmatrix}$$

Note that the inverse of the diagonal matrix Q, Q^{-1}, is another diagonal matrix

$$\begin{bmatrix} \dfrac{1}{Q_1} & & & & \\ & \dfrac{1}{Q_2} & & 0 & \\ & & \cdot & & \\ 0 & & & \cdot & \\ & & & & \dfrac{1}{Q_n} \end{bmatrix}$$

Using these matrices and the previously defined k, equations A43 become

$$Q^{-1}kQX = X\lambda$$

Rearranging in steps

$$kQX = QX\lambda$$
$$kQ = QX\lambda X^{-1}$$
$$k = QX\lambda X^{-1}Q^{-1} \tag{A44}$$

Let Δ be the determinant of X and Δ_{jl} be the cofactor of the jth row and the lth column of Δ. Then equation A44 can be multiplied out to yield the expression

$$k_{ij} = \frac{Q_i}{Q_j} \sum X_{il} \lambda_l \frac{\Delta_{jl}}{\Delta} \qquad (A45)$$

For each experiment n^2 values of X are determined, but because $a_i(0) = \sum X_{ij}$, only $n(n-1)$ of these are independent. However, there are also n values for λ so that the total number of independent values is n^2. Since the number of values of k to be determined in the system is also n^2 k is uniquely determined.

Thus, provided there is access to all the compartments and all the a can be determined as functions of time, only one experiment is required to determine all the values of X, λ, and Q. The values of k can then be calculated using equation A45. While dealing with simple cases in Chapter 3 it was possible to obtain expressions for X and λ in terms of the model parameters k, Q, etc. However, as the number of compartments in the model increases, the expressions for X and λ become impossibly complex, as can be seen by comparing the solutions for the one, two, and three compartment systems chosen. Indeed there is little value in knowing them or obtaining them. In the general case matrices have been used to calculate the values of k directly from the invariants X and λ. For all cases where n is greater than three it would seem to be easier to use matrices.

A4.3 Extension to open systems

A closed system can be converted to an open system by allowing each compartment in the system to exchange material with some undefined space outside the system. Compartment 1 of the general system of Figure A1 would then appear as in Figure A2. No matter to which compartments of the system tracer had been added, substance entering the system with flow rate R_{i0} $(i = 1, 2, \ldots n)$ would never contain tracer. However, tracer leaving the system with rate constants k_{0i} would contain tracer. As a result, the following modifications can be made to the equations of Section A4.2 in order to extend the discussion to open systems.

Equation A32 will become

$$\frac{d(Q_i)}{dt} = R_{i0} + \sum k_{ij}\,Q_j - Q_i \sum k_{li}$$

$$(i = 1, 2, \ldots n; \quad j = 1, 2, \ldots n; \quad l = 0, 1, \ldots n; j \neq i; l \neq i)$$

Equation A34 is now

$$\frac{d(q_i)}{dt} = \sum k_{ij}\,q_j - q_i \sum k_{li}$$

Equation A35 becomes

$$\frac{d(a_i)}{dt} = \frac{1}{Q_i} \sum k_{ij}\,a_j\,Q_j - a_i \sum k_{li}$$

In an open system tracer is lost from the system so that equation A36 is no longer true and is replaced by

$$\sum \frac{d(q_i)}{dt} \neq 0$$

The remainder of the argument of the previous Section still applies except that in the matrix $\mathbf{k}$

$$k_{ii} = \sum k_{li} \quad (l = 0, 1, \ldots n)$$

and as a consequence the first root λ_1 will not be zero.

In both open and closed systems, the solutions of the differential equations given by equation A42 has n terms in the summation. For an open system all of the terms are exponential terms, but for a closed system the first term is a constant. Thus the treatments in

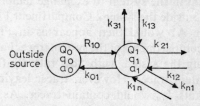

Fig. A2 A closed system is converted to an open system by introducing connections to some source outside the system.

the previous two Sections offer a mathematical proof for the general conclusions of Chapter 3.

A4.4 *Constrained systems*

If some of the elements of **k** in the general system are zero then the system is said to be constrained, a term first applied to compartmental analysis by Sheppard and Householder (1951). The more constraints that are applied to a system then the fewer parameters that need to be determined, and thus the less information that is required in order to determine these remaining parameters. It is sometimes possible to determine all the rate constants for a constrained system by making observations on fewer compartments.

Apart from some two compartment systems probably all the theoretical models of biological systems are constrained systems. Two common types of constrained systems will be considered i.e. mammillary and catenary systems.

A4.5 *Catenary systems*

An n-compartment catenary system was first defined by Sheppard and Householder (1951). This is a system in which the compartments are arranged in a chain, and each compartment only exchanges material with the one before and the one after. Figure A3 shows a closed n-compartment catenary system. The matrix in this case becomes

$$
\mathbf{k} = \begin{bmatrix}
-k_{11} & k_{12} & 0 & \ldots & 0 \\
k_{21} & -k_{22} & k_{23} & \ldots & 0 \\
0 & k_{32} & -k_{33} & \ldots & 0 \\
\multicolumn{5}{c}{\ldots\ldots\ldots\ldots\ldots\ldots\ldots\ldots\ldots} \\
0 & 0 & 0 & \ldots & -k_{nn}
\end{bmatrix}
$$

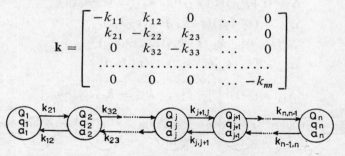

Fig. A3 A closed catenary system in which each compartment only exchanges material with the one before and the one after.

However, the commonest form of catenary system is the open system of Figure A4 where each compartment only transfers material to the compartment after it. $\mathbf{k}$ in this case is

$$\mathbf{k} = \begin{bmatrix} -k_{11} & 0 & 0 & \cdots & 0 \\ k_{21} & -k_{22} & 0 & \cdots & 0 \\ 0 & k_{32} & -k_{33} & \cdots & 0 \\ \cdots\cdots\cdots\cdots\cdots\cdots\cdots\cdots\cdots\cdots \\ 0 & 0 & 0 & \cdots & -k_{nn} \end{bmatrix}$$

For an open system, $k_{11} = k_{21}, k_{22} = k_{32}, \ldots k_{nn} = k_{0n}$ so that $s\mathbf{I} - \mathbf{k}$ becomes

$$\begin{bmatrix} (s+k_{21}) & 0 & 0 & \cdots & 0 \\ -k_{21} & (s+k_{32}) & 0 & \cdots & 0 \\ 0 & -k_{32} & (s+k_{43}) & \cdots & 0 \\ \cdots\cdots\cdots\cdots\cdots\cdots\cdots\cdots\cdots\cdots\cdots \\ 0 & 0 & 0 & \cdots & (s+k_{0n}) \end{bmatrix}$$

If tracer is injected into compartment 1 only (which is the usual case) then $a_1(0) = a_1(0)$ and $a_2(0) = a_3(0) = \cdots = 0$. Inserting these values into equation A40

$$A_j = \frac{\Delta_{1j}}{\Delta} \frac{Q_1}{Q_j} a_1(0)$$

Evaluating the determinants,

$$\Delta = (s+k_{21})(s+k_{32})(s+k_{43})\ldots(s+k_{0n})$$
$$\Delta_{11} = (s+k_{32})(s+k_{43})\ldots(s+k_{0n})$$
$$\Delta_{12} = -[-k_{21}(s+k_{43})\ldots(s+k_{0n})$$
$$\Delta_{13} = [k_{21}k_{32}(s+k_{54})\ldots(s+k_{0n})]$$
$$\cdots\cdots\cdots\cdots\cdots\cdots\cdots\cdots\cdots\cdots\cdots$$
$$\Delta_{1n} = (-1)^{1+n}[-k_{21}.-k_{32}.-k_{43}\ldots-k_{0n}]$$

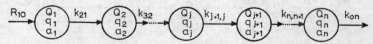

Fig. A4 The usual form for a catenary system. The system is open and material is only transported from one compartment to the succeeding one.

The expressions for A_j then become

$$A_1 = \frac{1}{(s+k_{21})} a_1(0)$$

$$A_2 = \frac{k_{21}}{(s+k_{21})(s+k_{32})} \frac{Q_1}{Q_2} a_1(0)$$

etc.

Hence the following important conclusion can be drawn. If compartment 1 is sampled, the concentration of tracer is described by one exponential term and from this k_{21} can be calculated. For compartment 2 the concentration would be described by two terms and k_{21} and k_{32} could be determined. In the general case, sampling the last compartment should allow all the rate constants of the model to be determined provided the curve analysis can be done satisfactorily.

A4.6 *Mammillary systems*

Sheppard and Householder (1951) were also the first to define an n-compartment mammillary system. This is a system containing a central compartment surrounded by $(n-1)$ peripheral compartments which exchange material with the central compartment

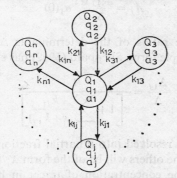

Fig. A5 A closed mammillary system consisting of a central compartment exchanging material with a series of peripheral compartments. The latter only exchange with the central compartment.

but not with each other. Figure A5 shows the theoretical model for an n-compartment closed mammillary system.

The matrices $\mathbf{A}$, $\mathbf{a}(0)$, $\mathbf{Q}$, and $\boldsymbol{\lambda}$ of the general system are unaltered, but $\mathbf{k}$ becomes

$$
\begin{bmatrix}
-k_{11} & k_{12} & k_{13} & \cdots & k_{1n} \\
k_{21} & -k_{22} & 0 & \cdots & 0 \\
k_{31} & 0 & -k_{33} & \cdots & 0 \\
\multicolumn{5}{c}{\dotfill} \\
k_{n1} & 0 & 0 & \cdots & -k_{nn}
\end{bmatrix}
$$

$\mathbf{sI} - \mathbf{k}$ becomes

$$
\begin{bmatrix}
(s+k_{11}) & -k_{12} & -k_{13} & \cdots & -k_{1n} \\
-k_{21} & (s+k_{22}) & 0 & \cdots & 0 \\
-k_{31} & 0 & (s+k_{33}) & \cdots & 0 \\
\multicolumn{5}{c}{\dotfill} \\
-k_{n1} & 0 & 0 & \cdots & (s+k_{nn})
\end{bmatrix}
$$

Tracer is usually injected into compartment 1 only, the central compartment, so that $a_1(0) = a_1(0)$ and $a_2(0) = a_3(0) = \cdots = a_n(0) = 0$. Inserting these values into equation A40 gives

$$
A_j = \frac{\Delta_{1j}}{\Delta} \frac{Q_1}{Q_j} a_1(0)
$$

For this case the ratios of the determinants Δ_{1j}/Δ cannot be expressed in a simple form. Sheppard (1962) has shown that

$$
\frac{\Delta_{1j}}{\Delta} = \frac{1}{s\left[1 + \sum \dfrac{k_{j1}}{(s+k_{1j})}\right]}
$$

so that A_j can be resolved into n partial fractions. One of these will be X_1/s and the others will be of the form $X_i/(s+\lambda_i)$. Thus the expressions for the concentration of tracer in both the central and the peripheral compartments contain a constant and $(n-1)$ exponential terms.

A4.7 Conclusion

The use of a general theory of compartmental analysis has found very little direct application to practical work, probably due to the introduction of computers allowing alternative methods to be used for analyzing experimental data. The rate constants, k_{ij}, can now be directly calculated from the experimental data and the mathematical model for the system, provided a suitable digital computer program is available (see Chapter 6). However, a knowledge of the general theory and its results can be very useful for obtaining a better understanding of the principles and use of compartmental analysis. It has therefore been presented in this Section for the benefit of those who wish to study the technique in greater depth.

REFERENCES

BERMAN, M. and SCHOENFELD, R. (1956). *J. appl. Phys.*, **27**, 1361. Invariants in experimental data on linear kinetics and the formulation of models.

HEARON, J. Z. (1953). *Bull. math. Biophys.*, **15**, 121. The kinetics of linear systems with special reference to periodic reactions.

SHEPPARD, C. W. (1948). *J. appl. Phys.*, **19**, 70. The theory of transfers within a multi-compartment system using isotopic tracers.

SHEPPARD, C. W. (1962). 'Basic Principles of the Tracer Method'. (John Wiley and Sons: New York.)

SHEPPARD, C. W. and HOUSEHOLDER, A. S. (1951). *J. appl. Phys.*, **22**, 510. The mathematical basis of the interpretation of tracer experiments in closed steady state systems.

TOBIAS, C. A. (1949). *Phys. Rev.*, **75**, 1460. Determination of the rate of biochemical reactions.

Index